CLEP* CALCULUS

Gregory Hill, M.Ed.

Mathematics Instructor
Hinsdale Central High School
Hinsdale, Illinois

Research & Education Association
Visit our website at: www.rea.com

Research & Education Association
61 Ethel Road West
Piscataway, New Jersey 08854
E-mail: info@rea.com

CLEP Calculus with Online Practice Exams

Published 2014

Copyright © 2013 by Research & Education Association, Inc.
Prior edition copyright © 2008. All rights reserved. No part of
this book may be reproduced in any form without permission of
the publisher.

Printed in the United States of America

Library of Congress Control Number 2012953356

ISBN-13: 978-0-7386-1101-3
ISBN-10: 0-7386-1101-8

All trademarks cited in this publication are the property of their respective owners.

LIMIT OF LIABILITY/DISCLAIMER OF WARRANTY: Publication of this work is for the purpose
of test preparation and related use and subjects as set forth herein. While every effort has been made
to achieve a work of high quality, neither Research & Education Association, Inc., nor the authors and
other contributors of this work guarantee the accuracy or completeness of or assume any liability in
connection with the information and opinions contained herein and in REA's software and/or online
materials. REA and the authors and other contributors shall in no event be liable for any personal
injury, property or other damages of any nature whatsoever, whether special, indirect, consequential
or compensatory, directly or indirectly resulting from the publication, use or reliance upon this work.

Cover image: © iStockphoto.com/mjbs

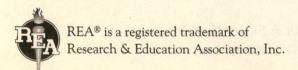

REA® is a registered trademark of
Research & Education Association, Inc.

CONTENTS

About Our Author vi

About Research & Education Association vi

Acknowledgments vi

CHAPTER 1

Passing the CLEP Calculus Exam 1

    Getting Started 3

    The REA Study Center 4

    An Overview of the Exam 5

    All About the CLEP Program 5

    Options for Military Personnel and Veterans 7

    SSD Accommodations for Candidates with Disabilities 8

    6-Week Study Plan 8

    Test-Taking Tips 9

    The Day of the Exam 10

CHAPTER 2

Limits and Continuity 11

    2.1   Introduction 13

    2.2   Limits as x Approaches a Constant 14

    2.3   Limits Involving Infinity 20

    2.4   Continuity 25

    2.5   Exercises 30

    2.6   Solutions to Exercises 32

CHAPTER 3

Concepts of the Derivative 37

    3.1   Introduction 39

    3.2   Rates of Change 39

    3.3   Introduction to Derivatives 44

    3.4   Linear Approximations 54

    3.5   Exercises 56

    3.6   Solutions to Exercises 58

CHAPTER 4

Rules of Differentiation **65**

   4.1 Introduction.. 67

   4.2 Derivatives of Polynomials 67

   4.3 Derivatives of Products and Quotients..................... 70

   4.4 Derivatives of Trigonometric Functions 74

   4.5 Derivatives of Exponential and Logarithmic Functions 78

   4.6 Higher-Order Derivatives... 81

   4.7 Chain Rule 81

   4.8 Implicit Differentiation 88

   4.9 Derivatives of Inverse Trigonometric Functions...... 91

  4.10 Derivatives of Inverse Functions............................. 94

  4.11 Exercises 97

  4.12 Solutions to Exercises... 99

CHAPTER 5

Applications of Differentiation **109**

   5.1 Introduction.. 111

   5.2 Functions and First-Derivative Applications 111

   5.3 Functions and Their Second-Derivative Applications 124

   5.4 Linear Particle Motion... 134

   5.5 Optimization 138

   5.6 Related Rates 143

   5.7 Linearization 146

   5.8 L'Hôpital's Rule 149

   5.9 Exercises 152

  5.10 Solutions to Exercises... 155

CHAPTER 6

Antidifferentiation and Definite Integrals................................ **165**

   6.1 Introduction.. 167

   6.2 Concept of the Antiderivative 167

   6.3 Numerical Approximation 175

   6.4 The Definite Integral as a Limit............................... 184

6.5   Properties of Definite Integrals.................................... 187
6.6   Fundamental Theorem of Calculus............................... 190
6.7   Exercises 197
6.8   Solutions to Exercises.. 200

CHAPTER 7

Applications of Integrals... **209**
7.1   Introduction.. 211
7.2   Differential Equations 211
7.3   Average Value 222
7.4   Area.. 227
7.5   Exercises 233
7.6   Solutions to Exercises.. 234

Practice Test 1 (also available online at *www.rea.com/studycenter*)............. **243**
Answer Key 262
Detailed Explanations of Answers....................................... 263

Practice Test 2 (also available online at *www.rea.com/studycenter*) **277**
Answer Key 297
Detailed Explanations of Answers....................................... 298

Answer Sheets... **313**

Glossary **371**

Index... **321**

ABOUT OUR AUTHOR

Greg Hill, a mathematics instructor at Hinsdale Central High School, Hinsdale, Ill., has taught all levels of mathematics for more than 25 years including more than 20 years of Advanced Placement Calculus. He received his B.A. in Math Education from the University of Illinois and his M.Ed. in Computer Education from National Louis University, Evanston, Ill. Mr. Hill and his wife, Cathy, have been married for 22 years and have two children, Andrea and Evan.

During his career, Mr. Hill has been a presenter at state, national and international mathematics conferences, has led summer technology workshops for educators, and since 1998 has facilitated summer AP Calculus institutes for fellow teachers. He has also had the opportunity to be a grader of AP Calculus exams. Among his professional recognitions, Mr. Hill was an Illinois state finalist for the Presidential Award for Excellence in Mathematics and Science Teaching in 1998 and 2000.

ABOUT RESEARCH & EDUCATION ASSOCIATION

Founded in 1959, Research & Education Association is dedicated to publishing the finest and most effective educational materials—including study guides and test preps—for students in middle school, high school, college, graduate school, and beyond.

Today, REA's wide-ranging catalog is a leading resource for teachers, students, and professionals. Visit *www.rea.com* to see a complete listing of our titles.

ACKNOWLEDGMENTS

In addition to our author, we would like to thank Larry B. Kling, Vice President, Editorial, for his overall guidance, which brought this publication to completion; Pam Weston, Publisher, for setting the quality standards for production integrity and managing the publication to completion; Diane Goldschmidt, Senior Editor, for editorial project management; Alice Leonard and Mike Reynolds, Senior Editors, for preflight editorial review; Christine Saul, Senior Graphic Designer, for designing our cover; Sandra Rush for copyediting the manuscript; Mel Friedman for his technical review of the manuscript; and Kathy Caratozzolo for typesetting this edition.

CHAPTER 1

Passing the
CLEP Calculus Exam

PASSING THE CLEP CALCULUS EXAM

Congratulations! You're joining the millions of people who have discovered the value and educational advantage offered by the College Board's College-Level Examination Program, or CLEP. This test prep focuses on what you need to know to succeed on the CLEP Calculus exam, and will help you earn the college credit you deserve while reducing your tuition costs.

GETTING STARTED

There are many different ways to prepare for a CLEP exam. What's best for you depends on how much time you have to study and how comfortable you are with the subject matter. To score your highest, you need a system that can be customized to fit you: your schedule, your learning style, and your current level of knowledge.

This book, and the online tools that come with it, allow you to create a personalized study plan through three simple steps: assessment of your knowledge, targeted review of exam content, and reinforcement in the areas where you need the most help.

Let's get started and see how this system works.

Test Yourself & Get Feedback	Assess your strengths and weaknesses. The score report from your online diagnostic exam gives you a fast way to pinpoint what you already know and where you need to spend more time studying.
Review with the Book	Armed with your diagnostic score report, review the parts of the book where you're weak and study the answer explanations for the test questions you answered incorrectly.
Ensure You're Ready for Test Day	After you've finished reviewing with the book, take our full-length practice tests. Review your score reports and re-study any topics you missed. We give you two full-length practice tests to ensure you're confident and ready for test day.

THE REA STUDY CENTER

The best way to personalize your study plan is to get feedback on what you know and what you don't know. At the online REA Study Center, you can access two types of assessment: a diagnostic exam and full-length practice exams. Each of these tools provides true-to-format questions and delivers a detailed score report that follows the topics set by the College Board.

Diagnostic Exam

Before you begin your review with the book, take the online diagnostic exam. Use your score report to help evaluate your overall understanding of the subject, so you can focus your study on the topics where you need the most review.

Full-Length Practice Exams

These practice tests give you the most complete picture of your strengths and weaknesses. After you've finished reviewing with the book, test what you've learned by taking the first of the two online practice exams. Review your score report, then go back and study any topics you missed. Take the second practice test to ensure you have mastered the material and are ready for test day.

If you're studying and don't have Internet access, you can take the printed tests in the book. These are the same practice tests offered at the REA Study Center, but without the added benefits of timed testing conditions and diagnostic score reports. Because the actual exam is Internet-based, we recommend you take at least one practice test online to simulate test-day conditions.

AN OVERVIEW OF THE EXAM

The CLEP Calculus exam consists of approximately 44 multiple-choice questions, each with five possible answer choices, to be answered in 90 minutes.

The exam covers the material one would find in a college-level Calculus course. The exam focuses on the test-taker's understanding of calculus and his/her experience with its methods and applications. The test-taker's knowledge of other areas of mathematics, including algebra, plane and solid geometry, trigonometry and analytic geometry is assumed. Use of a graphing calculator (non-CAS) is allowed ONLY in Section 2 of the exam.

The approximate breakdown of topics is as follows:

60%   Limits and Differential Calculus

40%   Integral Calculus

ALL ABOUT THE CLEP PROGRAM

What is the CLEP?

CLEP is the most widely accepted credit-by-examination program in North America. The CLEP program's 33 exams span five subject areas. The exams assess the material commonly required in an introductory-level college course. Examinees can earn from three to twelve credits at more than 2,900 colleges and universities in the U.S. and Canada. For a complete list of the CLEP subject examinations offered, visit the College Board website: *www.collegeboard.org/clep*.

Who takes CLEP exams?

CLEP exams are typically taken by people who have acquired knowledge outside the classroom and who wish to bypass certain college courses and earn college credit. The CLEP program is designed to reward examinees for learning—no matter where or how that knowledge was acquired.

Although most CLEP examinees are adults returning to college, many graduating high school seniors, enrolled college students, military personnel, veterans, and international students take CLEP exams to earn college credit or to demonstrate their ability to perform at the college level. There are no prerequisites, such as age or educational status, for taking CLEP examinations. However, because policies on granting credits vary among colleges, you should contact the particular institution from which you wish to receive CLEP credit.

How is my CLEP score determined?

Your CLEP score is based on two calculations. First, your CLEP raw score is figured; this is just the total number of test items you answer correctly. After the test is administered, your raw score is converted to a scaled score through a process called *equating*. Equating adjusts for minor variations in difficulty across test forms and among test items, and ensures that your score accurately represents your performance on the exam regardless of when or where you take it, or on how well others perform on the same test form.

Your scaled score is the number your college will use to determine if you've performed well enough to earn college credit. Scaled scores for the CLEP exams are delivered on a 20–80 scale. Institutions can set their own scores for granting college credit, but a good passing estimate (based on recommendations from the American Council on Education) is generally a scaled score of 50, which usually requires getting roughly 66% of the questions correct.

For more information on scoring, contact the institution where you wish to be awarded the credit.

Who administers the exam?

CLEP exams are developed by the College Board, administered by Educational Testing Service (ETS), and involve the assistance of educators from throughout the United States. The test development process is designed and implemented to ensure that the content and difficulty level of the test are appropriate.

When and where is the exam given?

CLEP exams are administered year-round at more than 1,200 test centers in the United States and can be arranged for candidates abroad on request. To find the test center nearest you and to register for the exam, contact the CLEP Program:

CLEP Services
P.O. Box 6600
Princeton, NJ 08541-6600
Phone: (800) 257-9558 (8 a.m. to 6 p.m. ET)
Fax: (610) 628-3726
Website: *www.collegeboard.org/clep*

CLEP exams migrating to iBT

To improve the testing experience for both institutions and test-takers, the College Board's CLEP Program is transitioning its 33 exams from the eCBT platform to an Internet-based testing (iBT) platform. By spring 2014, all CLEP test-takers will be able to register for exams and manage their personal account information through the "My Account" feature on the CLEP website. This new feature simplifies the registration process and automatically downloads all pertinent information about the test session, making for a more streamlined check-in.

OPTIONS FOR MILITARY PERSONNEL AND VETERANS

CLEP exams are available free of charge to eligible military personnel and eligible civilian employees. All the CLEP exams are available at test centers on college campuses and military bases. Contact your Educational Services Officer or Navy College Education Specialist for more information. Visit the DANTES or College Board websites for details about CLEP opportunities for military personnel.

Eligible U.S. veterans can claim reimbursement for CLEP exams and administration fees pursuant to provisions of the Veterans Benefits Improvement Act of 2004. For details on eligibility and submitting a claim for reimbursement, visit the U.S. Department of Veterans Affairs website at *www.gibill.va.gov.*

CLEP can be used in conjunction with the Post-9/11 GI Bill, which applies to veterans returning from the Iraq and Afghanistan theaters of operation. Because the GI Bill provides tuition for up to 36 months, earning college credits with CLEP exams expedites academic progress and degree completion within the funded timeframe.

SSD ACCOMMODATIONS FOR CANDIDATES WITH DISABILITIES

Many test candidates qualify for extra time to take the CLEP exams, but you must make these arrangements in advance. For information, contact:

College Board Services for Students with Disabilities
P.O. Box 8060
Mt. Vernon, IL 62864-0060
Phone: (609) 771-7137 (Monday through Friday, 8 a.m. to 6 p.m. ET)
TTY: (609) 882-4118
Fax: (866) 360-0114
E-mail: ssd@info.collegeboard.org

6-WEEK STUDY PLAN

Although our study plan is designed to be used in the six weeks before your exam, it can be condensed to three weeks by combining each two-week period into one.

Be sure to set aside enough time—at least two hours each day—to study. The more time you spend studying, the more prepared and relaxed you will feel on the day of the exam.

Week	Activity
1	Take the Diagnostic Exam. The score report will identify topics where you need the most review.
2–4	Study the review chapters. Use your diagnostic score report to focus your study.
5	Take Practice Test 1 at the REA Study Center. Review your score report and re-study any topics you missed.
6	Take Practice Test 2 at the REA Study Center to see how much your score has improved. If you still got a few questions wrong, go back to the review and study any topics you may have missed.

TEST-TAKING TIPS

Know the format of the test. Familiarize yourself with the CLEP computer screen beforehand by logging on to the College Board website. Waiting until test day to see what it looks like in the pretest tutorial risks injecting needless anxiety into your testing experience. Also, familiarizing yourself with the directions and format of the exam will save you valuable time on the day of the actual test.

Read all the questions—completely. Make sure you understand each question before looking for the right answer. Reread the question if it doesn't make sense.

Read all of the answers to a question. Just because you think you found the correct response right away, do not assume that it's the best answer. The last answer choice might be the correct answer.

Work quickly and steadily. You will have 90 minutes to answer 44 questions, so work quickly and steadily. Taking the timed practice tests online will help you learn how to budget your time.

Use the process of elimination. Stumped by a question? Don't make a random guess. Eliminate as many of the answer choices as possible. By eliminating just two answer choices, you give yourself a better chance of getting the item correct, since there will only be three choices left from which to make your guess. Remember, your score is based only on the number of questions you answer correctly.

Don't waste time! Don't spend too much time on any one question. Remember, your time is limited and pacing yourself is very important. Work on the easier questions first. Skip the difficult questions and go back to them if you have the time.

Look for clues to answers in other questions. If you skip a question you don't know the answer to, you might find a clue to the answer elsewhere on the test.

Be sure that your answer registers before you go to the next item. Look at the screen to see that your mouse-click causes the pointer to darken the proper oval. If your answer doesn't register, you won't get credit for that question.

THE DAY OF THE EXAM

On test day, you should wake up early (after a good night's rest, of course) and have breakfast. Dress comfortably, so you are not distracted by being too hot or too cold while taking the test. (Note that "hoodies" are not allowed.) Arrive at the test center early. This will allow you to collect your thoughts and relax before the test, and it will also spare you the anxiety that comes with being late.

Before you leave for the test center, make sure you have your admission form and another form of identification, which must contain a recent photograph, your name, and signature (i.e., driver's license, student identification card, or current alien registration card). You may wear a watch. However, you may not wear one that makes noise, because it may disturb the other test-takers. No cell phones, dictionaries, textbooks, notebooks, briefcases, or packages will be permitted, and drinking, smoking, and eating are prohibited.

Good luck on the CLEP Calculus exam!

CHAPTER 2

Limits and Continuity

CHAPTER 2

LIMITS AND CONTINUITY

2.1   INTRODUCTION

Limits are the building blocks of calculus. They are used to establish all major concepts, including continuity, derivatives, and integrals. There are many ways to examine and evaluate limits. Some of the methods that we will review are studying numerical patterns, direct substitution, deducing information from a graph, simplifying prior to substitution, and taking an intuitive approach to the behavior of a particular function. Limits are also closely linked with continuity. Prior to establishing a formal definition of continuity, you should recall from previous courses that any kind of a break in the domain of a function is called a discontinuity. Whether a function is continuous or discontinuous at a point in its domain determines the ease with which a limit at that point may be determined.

Imagine a square with an area of 4 square feet (Figure 2.1). If the consecutive midpoints of each side of the square are connected with segments, a new square is formed with an area of 2 square feet. If the consecutive midpoints of the new square are connected, the resulting area is 1 square foot. Repeating this pattern over and over, the sequence of areas is 4, 2, 1, $\frac{1}{2}$, $\frac{1}{4}$, ... Even though each new area could continue to be multiplied by $\frac{1}{2}$ forever, it can be said that the limit of the areas is 0. Interestingly, you

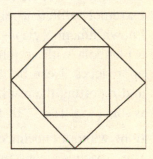

Figure 2.1

may recognize this sequence from a previous course as an infinite geometric sequence with a common ratio of $\frac{1}{2}$. The sum of the infinite number of consecutive areas has a limit given by the formula $S_\infty = \dfrac{a}{1-r}$, where a is the first term, and r is the common ratio. In this case, $S_\infty = \dfrac{4}{1-\frac{1}{2}} = 8$ square feet. If n is the number of squares whose areas are being summed, the previous conclusion could also be written as $\lim\limits_{n\to\infty} S_n = 8$.

2.2   LIMITS AS X APPROACHES A CONSTANT

A NUMERICAL APPROACH

Table 2.1

x	2.97	2.98	2.99	3.01	3.02	3.03
$f(x)$	4.8209	4.8804	4.9401	5.0601	5.1204	5.1809

Table 2.1 lists ordered pairs for a certain function, $f(x)$. Examining the table might lead one to believe that the function has a certain limit as x approaches 3. Without knowing the actual function, we might speculate that the limit is 5, but lacking more information, there is no guarantee that for x values even closer to $x = 3$, the function won't behave erratically and "zoom" off to some extremely different value. Actually, the values given were generated by a known function, $f(x) = x^2 - 4$. This, of course, is a "well-behaved" function— a parabola with no jumps or breaks in its domain. As a result, its limit as x approaches 3 can be determined by simply substituting the value of 3 for x and reporting the resulting function value, in this case, 5. The proper notation is $\lim\limits_{x\to 3}(x^2 - 4) = 5$. In fact, for every polynomial function and all other functions without undefined points, or discontinuities, in their domain, the limit as x approaches a constant c, can be evaluated by direct substitution. The limit is equal to $f(c)$ and is written as $\lim\limits_{x\to c} f(x) = f(c)$.

A GRAPHICAL APPROACH

Since many functions are not defined over all real numbers, it is important to know how to deal with limits as x approaches points where there are missing values of the domain. To do so, one must understand one-sided limits.

A general limit exists at a point $x = c$ when the limit as x approaches c from the left equals the limit as x approaches c from the right. The actual function value at $x = c$ does not need to exist. In symbols, the limit as x approaches c from the left is denoted $\lim_{x \to c^-} f(x)$. The negative sign above and to the right of the c can be thought of as telling us to examine function values associated with x values less than c. Likewise, the limit as x approaches c from the right is denoted $\lim_{x \to c^+} f(x)$.

Existence of a Limit

If $f(x)$ is a function, and c and L are real constants, then for any interior point of the domain of f, $\lim_{x \to c} f(x) = L$ if and only if $\lim_{x \to c^-} f(x) = \lim_{x \to c^+} f(x) = L$.

If a function is defined on an open, closed, or half-open interval, at either endpoint of the domain, only the appropriate one-sided limit is examined.

EXAMPLE 2.1

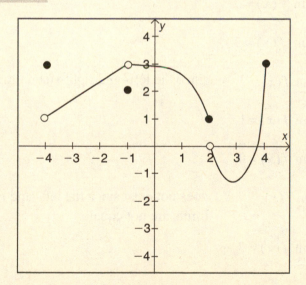

Figure 2.2

The graph of the function $f(x)$ shown in Figure 2.2 is defined only on the domain shown. Determine the following:

(A) $\quad \lim_{x \to -4^+} f(x)$

(B) $\quad \lim_{x \to -1^-} f(x)$

(C) $\quad \lim_{x \to -1^+} f(x)$

(D) $\quad \lim_{x \to -1} f(x)$

(E) $\quad \lim_{x \to 2^+} f(x)$

(F) $\quad \lim_{x \to 2^+} f(x)$

(G) $\quad \lim_{x \to 2} f(x)$

(H) $\quad \lim_{x \to 4^-} f(x)$

SOLUTION

(A) $\quad \lim_{x \to -4^+} f(x) = 1$

(B) $\quad \lim_{x \to -1^-} f(x) = 3$

(C) $\quad \lim_{x \to -1^+} f(x) = 3$

(D) $\quad \lim_{x \to -1} f(x) = 3 \qquad$ since the left- and right-sided limits are equal.

(E) $\quad \lim_{x \to 2^-} f(x) = 1$

(F) $\quad \lim_{x \to 2^+} f(x) = 0$

(G) $\quad \lim_{x \to 2} f(x) \qquad$ does not exist since the left- and right-sided limits are not equal.

(H) $\quad \lim_{x \to 4^-} f(x) = 3$

EXAMPLE 2.2

Find $\lim\limits_{x\to 0}\dfrac{\sin(x)}{x}$ by examining its graph (Figure 2.3).

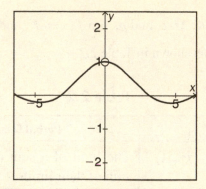

Figure 2.3

SOLUTION

The function is not defined at $x = 0$, but the y-values are clearly approaching 1, therefore $\lim\limits_{x\to 0}\dfrac{\sin(x)}{x}$. A more convincing argument comes from the Sandwich Theorem.

The Sandwich Theorem

If $f(x) \le g(x) \le h(x)$ for x near some constant c, and $\lim\limits_{x\to c}f(x)=\lim\limits_{x\to c}h(x)=L$, then $\lim\limits_{x\to c}g(x)=L$.

It can be shown that for x near 0, $\cos(x)\le\dfrac{\sin(x)}{x}\le 1$, so $\lim\limits_{x\to 0}\cos(x)\le$ $\lim\limits_{x\to 0}\dfrac{\sin(x)}{x}\le\lim\limits_{x\to 0}1$. $\lim\limits_{x\to 0}\cos(x)=\lim\limits_{x\to 0}1=1$, so $\lim\limits_{x\to 0}\dfrac{\sin(x)}{x}=1$. Presented without proof, more generally, for nonzero constants a and b, $\lim\limits_{x\to 0}\dfrac{\sin(ax)}{bx}=\dfrac{a}{b}$.

PROPERTIES OF LIMITS

In addition, many properties of limits can be utilized to evaluate limits of more complicated or unfamiliar functions.

Given real numbers K, M, c, and p, $\lim\limits_{x \to c} f(x) = K$, and $\lim\limits_{x \to c} h(x) = M$, then the properties of limits are as shown in Table 2.2.

Table 2.2

Property	Verbal Description
$\lim\limits_{x \to c}[f(x) + h(x)] = K + M$	The limit of a sum of functions is the sum of their limits.
$\lim\limits_{x \to c}[f(x) - h(x)] = K - M$	The limit of a difference of functions is the difference of their limits.
$\lim\limits_{x \to c}[p \cdot h(x)] = p \cdot M$	The limit of a constant times a function is the constant times the limit.
$\lim\limits_{x \to c}[f(x) \cdot h(x)] = K \cdot M$	The limit of a product of functions is the product of their limits.
$\lim\limits_{x \to c}\left[\dfrac{f(x)}{h(x)}\right] = \dfrac{K}{M},\ M \neq 0$	The limit of a quotient of functions is the quotient of their limits, provided the denominator is not 0.

AN ALGEBRAIC APPROACH

As stated previously, evaluating a limit for a continuous function can be done by simple substitution. Even though all of the steps are rarely shown, many of the properties of limits are actually being applied.

EXAMPLE 2.3

Find $\lim\limits_{x \to 5}(x^2 + 3x - 4)$.

SOLUTION

$$\lim_{x \to 5} (x^2 + 3x - 4) = \lim_{x \to 5} x^2 + \lim_{x \to 5} 3x - \lim_{x \to 5} 4)$$

$$= (\lim_{x \to 5} x)^2 + 3 \lim_{x \to 5} x - \lim_{x \to 5} 4$$

$$= 5^2 + 3 \cdot 5 - 4$$

$$= 36$$

Sometimes direct substitution is not possible, however, and manipulating the form of the function may be necessary prior to evaluating the limit. When trying to find $\lim_{x \to 2} \dfrac{x^2 - 4}{x - 2}$, for example, substituting a 2 for x would result in the undefined ratio $\dfrac{0}{0}$. In this case, factoring and reducing the function prior to substitution is the key.

EXAMPLE 2.4

Find $\lim_{x \to 2} \dfrac{x^2 - 4}{x - 2}$.

SOLUTION

$$\lim_{x \to 2} \frac{x^2 - 4}{x - 2} = \lim_{x \to 2} \frac{(x - 2)(x + 2)}{x - 2}$$

$$= \lim_{x \to 2} (x + 2)$$

$$= 2 + 2$$

$$= 4$$

This limit can also be confirmed numerically by examining a table of values near $x = 2$, such as Table 2.3.

Table 2.3

x	1.997	1.998	1.999	2	2.001	2.002	2.003
$\dfrac{x^2 - 4}{x - 2}$	3.997	3.998	3.999	Error	4.001	4.002	4.003

As required, the domain values from the left and the right of $x = 2$ are producing function values approaching 4, even though the function is not defined at $x = 2$.

Finally, thinking graphically, you should recall from previous courses that a factor completely canceled from the denominator of a rational expression leaves a hole in the graph at the zero of that factor. Thus, the graph of $f(x) = \dfrac{x^2 - 4}{x - 2}$ looks exactly like the graph of $y = x + 2$ except that $f(x)$ has a hole in the graph at the point (2, 4), since 2 is the zero of the canceled factor $x - 2$.

2.3 LIMITS INVOLVING INFINITY

AN INFINITE RESULT AS X APPROACHES A CONSTANT

Sometimes a rational expression cannot be manipulated to make direct evaluation possible. Often, the result is a ratio that grows without bound. For example, given $f(x) = \dfrac{4}{x - 3}$, find $\displaystyle\lim_{x \to 3^+} = \dfrac{4}{x - 3}$. Substituting 3 for x would result in division by 0. Also, $x - 3$ cannot be canceled from the denominator. In cases such as this, examining the function numerically is useful. As x gets closer and closer to 3 from the right, the denominator will get smaller and smaller, but remain positive. Dividing 4 by smaller and smaller positive numbers will result in ever-increasing ratios, as shown in Table 2.4.

Table 2.4

x	3.1	3.01	3.001	3.0001
$\dfrac{4}{x-3}$	40	400	4000	40000

You may also remember that the graph of $f(x) = \dfrac{4}{x - 3}$ has a vertical asymptote at $x = 3$. This gives visual confirmation that $\displaystyle\lim_{x \to 3^+} \dfrac{4}{x - 3} = \infty$. (*Note:*

Infinity is not a real number. The notation simply implies that the value of $\frac{4}{x-3}$ is unbounded.)

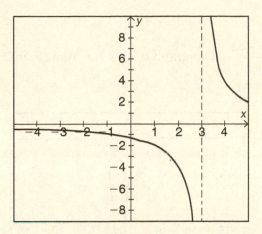

Figure 2.4

The graph in Figure 2.4 also sheds some light on why a one-sided limit was examined. Since the function values go to ∞ on the right of $x = 3$, and $-\infty$ on the left of $x = 3$, the limit as x approaches 3 does not exist. (DNE, a common shorthand notation for "does not exist," will be used throughout this text when appropriate.) Remember, for a limit to exist at an interior point of a domain, the left- and right-sided limits must exist and be equal.

LIMITS AS *X* APPROACHES POSITIVE OR NEGATIVE INFINITY

It can also be useful to examine the behavior of functions when the absolute value of x gets infinitely large. In preparatory courses for calculus, this is called "examining end behavior." It is a method intended to determine how a function is behaving "way out on the ends." Only a few different things can happen to the function values: they may approach a constant, their absolute values may grow without bound, they may oscillate, or they may behave erratically. A limit, as x approaches positive or negative infinity, that goes toward a constant indicates the presence of a horizontal asymptote.

EXAMPLE 2.5

Find any horizontal asymptotes of $g(x) = \dfrac{2x^4 - 5x}{3 + x^4}$.

SOLUTION

Let $g(x) = \dfrac{2x^4 - 5x}{3 + x^4}$. The graph of this function (Figure 2.5) shows $g(x)$ on the left and right ends.

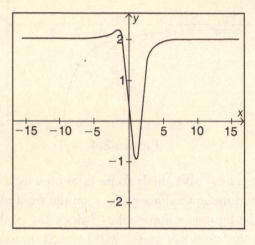

Figure 2.5

It appears to be leveling off to $y = 2$, as the table of values (Table 2.5) also indicates, even though the values of x are relatively very small.

Table 2.5

x	-100	-20	-15	-10	10	15	50	100
$g(x)$	2.0000049	2.00058	2.00136	2.00439	1.9944	1.9984	1.999959	1.999994

Analytically, one can examine the limit by dividing each term by the highest power of the independent variable.

$$g(x) = \frac{2x^4 - 5x}{3 + x^4} \cdot \frac{\frac{1}{x^4}}{\frac{1}{x^4}}$$

$$g(x) = \frac{2 - \frac{5}{x^3}}{\frac{3}{x^4} + 1}$$

$$\lim_{x\to\infty} = \frac{2 - \frac{5}{x^3}}{\frac{3}{x^4} + 1} = \frac{2-0}{0+1} = 2$$

As x gets infinitely large, $\dfrac{5}{x^3}$ and $\dfrac{3}{x^4}$ each get infinitely small and the ratio approaches 2. The horizontal asymptote of $g(x) = \dfrac{2x^4 - 5x}{3 + x^4}$ is the line $y = 2$.

EXAMPLE 2.6

Find any horizontal asymptotes of $h(x) = \dfrac{x^2 - 2x}{5x + 9}$.

SOLUTION

A numerical or graphical examination of $h(x)$ would indicate $\lim\limits_{x\to\infty} h(x) = \infty$. This notation simply means the function values grow without bound. We must not think of ∞ as a number. There is a logical reason to expect this result. Consider $h(1000)$. The numerator will be close to 1,000,000, but the denominator will be only around 5000. As x gets even larger, the squaring of an extremely large x value will far outgrow $5x$. Also, in the numerator, for very large values of x, $2x$ will become insignificant in size relative to x^2. In the denominator, 9 will become insignificant relative to the value of $5x$.

You may remember this idea from a previous course. Examining the ratio of just the highest degree terms of the numerator and denominator provides an end behavior model that gives insight into limits at positive and negative infinity.

In general, for rational functions of the form $y = \dfrac{f(x)}{g(x)}$, where f and g are polynomials, the relative degree of each polynomial will determine the limit as x gets infinitely positive or negative.

Degree of $g(x) >$ Degree of $f(x) \Rightarrow \lim\limits_{x\to\infty} \dfrac{f(x)}{g(x)} = 0$

Degree of $g(x) <$ Degree of $f(x) \Rightarrow \lim\limits_{x\to\infty} \dfrac{f(x)}{g(x)} = \infty$

Degree of $g(x) =$ Degree of $f(x) \Rightarrow \lim\limits_{x\to\infty} \dfrac{f(x)}{g(x)} = c$, where c is the ratio of the leading coefficients of f and g.

EXAMPLE 2.7

Find $\lim\limits_{x \to \infty} \dfrac{7x+3+5x^2}{9x^2-1}$.

SOLUTION

$\lim\limits_{x \to \infty} \dfrac{7x+3+5x^2}{9x^2-1} = \dfrac{5}{9}$. Since the degree of the numerator and denominator are equal, the limit is the ratio of the leading coefficients, $\dfrac{5}{9}$. For extremely large values of x, the other terms become insignificant.

EXAMPLE 2.8

Find $\lim\limits_{x \to -\infty} \dfrac{3x-7}{x^2+11}$.

SOLUTION

$\lim\limits_{x \to -\infty} \dfrac{3x-7}{x^2+11} = 0$ since the degree of the denominator is larger than the degree of the numerator.

Of course, not all functions are rational functions, so we must be ready to explore limits by using all three methods: analytical, numerical, and graphical.

EXAMPLE 2.9

Find $\lim\limits_{x \to \infty} x^2 - 3x$.

SOLUTION

One might consider thinking of the problem as $\lim\limits_{x \to \infty} x^2 - 3x = \lim\limits_{x \to \infty} x^2 - \lim\limits_{x \to \infty} 3x$. Even though this takes on a form of $\infty - \infty$, the limit is not 0. It can be solved easily by thinking about the graph of the function. The graph is a parabola that opens upward. As x gets infinitely large, y gets infinitely large.

$$\lim\limits_{x \to \infty} x^2 - 3x = \infty$$

Saying "the limit does not exist" is also acceptable.

Another method of evaluating limits is a substitution for $\dfrac{1}{x}$. This will turn a limit with x approaching infinity into a limit approaching 0.

EXAMPLE 2.10

Find $\lim\limits_{x \to \infty}\left[\dfrac{1}{4}\, x \cdot \sin\left(\dfrac{3}{x}\right)\right]$.

SOLUTION

Let $a = \dfrac{1}{x}$, so $x = \dfrac{1}{a}$, and as $x \to \infty$, $a \to 0$ from the right.

$$\lim_{x \to \infty}\left[\frac{1}{4}\, x \cdot \sin\left(\frac{3}{x}\right)\right] = \lim_{a \to 0^+} \frac{1}{4}\,\frac{1}{a}\,\sin(3a)$$

$$= \lim_{a \to 0^+} \frac{\sin(3a)}{4a}$$

$$= \frac{3}{4}$$

Notice the use of the generalized form $\lim\limits_{x \to 0} \dfrac{\sin(ax)}{bx} = \dfrac{a}{b}$.

In general, when evaluating limits, be aware of and open to using any of the available methods.

2.4   CONTINUITY

Continuity is another critical building block for calculus. Once the idea of continuity is established, almost every major theorem has the condition of continuity at a point or on an interval included in the hypothesis.

CONTINUITY AT A POINT

A function is continuous at a point $x = c$ if and only if $\lim\limits_{x \to c} f(x) = f(c)$.

This statement has three implicit conditions:

1. $\lim\limits_{x \to c} f(x)$ exists;

2. $f(c)$ exists; and

3. $\lim\limits_{x \to c} f(x)$ and $f(c)$ have the same value.

A function $f(x)$ may be discontinuous at a point if any one of the three conditions fails to hold true. Logically, if one of the first two conditions fails, the third condition will also fail.

Figures 2.6–2.8 show possible scenarios for discontinuity of a function $f(x)$. Common types of discontinuities include "jump," as shown in Figure 2.6; "removable," as shown in Figure 2.7 and Figure 2.8; and vertical asymptotes, sometimes called "infinite discontinuities," which occur in many rational functions. Removable discontinuities occur commonly when a factor cancels completely from the denominator of a rational function. When a factor does not cancel completely from the denominator, the zero of the remaining factor determines the location of a vertical asymptote.

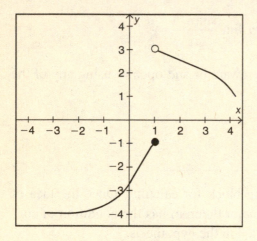

Figure 2.6

$\lim\limits_{x \to 1} f(x)$ does not exist

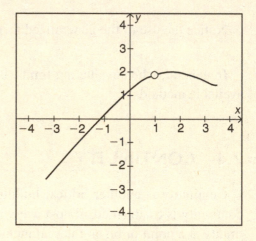

Figure 2.7

$f(1)$ does not exist,
but $\lim\limits_{x \to 1} f(x)$ does exist

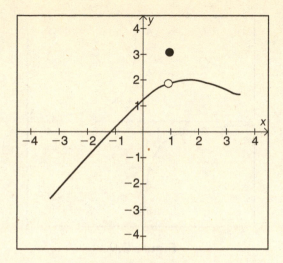

Figure 2.8

$$\lim_{x \to 1} f(x) \neq f(1)$$

EXAMPLE 2.11

Find all discontinuities of $g(x) = \dfrac{x-4}{x^2 - 4x}$ over the real numbers.

SOLUTION

$$g(x) = \frac{x-4}{x(x-4)}$$

$$= \frac{1}{x}$$

Since $x - 4$ canceled completely, the graph has a removable discontinuity at $x = 4$. Since x remains as a factor in the denominator, its zero, $x = 0$ (the y-axis), is a vertical asymptote. The graph of the function in Figure 2.9 provides visual confirmation.

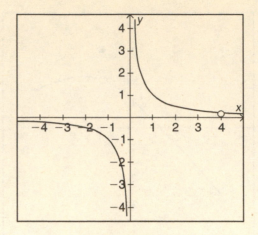

Figure 2.9

A common task is to redefine a function to "remove" the discontinuity at a certain point. This can be done only when the discontinuity is indeed removable! In the preceding example, $\lim\limits_{x \to 4} g(x) = \dfrac{1}{4}$. This means the "hole" in the graph lies at $\left(4, \dfrac{1}{4}\right)$. $g(x)$ can then be redefined as the piecewise function, with continuity at $x = 4$,

$$g(x) = \begin{cases} \frac{x-4}{x^2-4x}, & x \neq 4 \\ \frac{1}{4}, & x = 4 \end{cases}$$

Taking the limit of $g(x)$ as x goes to positive or negative infinity also reveals the horizontal asymptote. Because the denominator has a higher degree than the numerator, $\lim\limits_{x \to \infty} \dfrac{x-4}{x^2-4x} = 0$. The equation of the horizontal asymptote is $y = 0$.

Finally, note that there exist many functions besides rational functions that have removable, infinite, or jump discontinuities. For example, $y = \dfrac{\sin(x)}{x}$ is a different kind of function with a removable discontinuity. You should remember that $y = \ln(x)$, $y = \tan(x)$, and $y = \sec(x)$ all have vertical asymptotes. In addition, the greatest integer function has jump discontinuities at every integer in its domain.

CONTINUITY OF A FUNCTION

In the absence of a stated domain, a function is continuous if it is continuous at every point of its domain. Interestingly, this means that, for instance,

$f(x) = \dfrac{1}{x}$ is considered a continuous function since $x = 0$ is not in its domain, even though most people would see it as being discontinuous at its vertical asymptote. Fortunately, this approach to continuity of functions is rarely used in introductory calculus courses. Most often, one is asked to determine continuity over a stated domain, or determine points of discontinuity over the real numbers. In this context, $f(x) = \dfrac{1}{x}$ is not continuous *over the real numbers*.

EXAMPLE 2.12

On the domain $[-4, 4]$, where is $f(x) = \dfrac{2}{x^2 - 4}$ discontinuous?

SOLUTION

$$f(x) = \frac{2}{x^2 - 4}$$

$$= \frac{2}{(x-2)(x+2)}$$

Zeros in the denominator cause discontinuities, so $f(x)$ is discontinuous at $x = 2$ and $x = -2$.

It is also important to note that any algebraic combination of continuous functions is also a continuous function with the exception of division by a function with a value of 0.

If g and h are continuous at a, then the following algebraic combinations of functions are continuous at a:

1) $g \pm h$ 2) $g \cdot h$ 3) $\dfrac{g}{h}$ if $h(a) \neq 0$

Additionally, a composition of continuous functions is also continuous.

If g is continuous at a, and h is continuous at $g(a)$, then $h(g(x))$ is also continuous at a.

One of the most important immediate consequences of continuity is the Intermediate Value Theorem. We state it without proof.

Intermediate Value Theorem

> Let f be continuous on the closed interval $[a, b]$, and let M be any number between $f(a)$ and $f(b)$. There exists a number c in (a, b) such that $f(c) = M$.

Informally, this theorem says that on a given interval $[a, b]$, a continuous function will, at least once, take on all y-values between $f(a)$ and $f(b)$. One important use of the Intermediate Value Theorem is justifying the existence of roots of equations.

EXAMPLE 2.13

Show that a root of the equation $x^3 - 3x^2 - 2x + 1 = 0$ lies between $x = 0$ and $x = 1$.

SOLUTION

Let $f(x) = x^3 - 3x^2 - 2x + 1$.

Since $f(0) = 1$ and $f(1) = -3$, and all polynomial functions are continuous, function f will take on all values between -3 and 1 at least once in the given interval $(0, 1)$. By the Intermediate Value Theorem, there exists a c in the interval $(0, 1)$ such that $f(c) = 0$.

2.5 EXERCISES

Work the following exercises without a calculator. Solutions are given in the next section.

Evaluate the following limits by the method of your choice.

1. $\displaystyle\lim_{x \to -5}(x^2 - 9x)$

2. $\displaystyle\lim_{x \to 3} \frac{x^3 - 9x}{x - 3}$

3. $\displaystyle\lim_{x \to 0} \frac{\sin(5x)}{8x}$

4. $\lim\limits_{x \to 7^+} \dfrac{2}{x-7}$

5. $\lim\limits_{x \to \infty} 3^{\left(\frac{1}{x}\right)}$

6. $\lim\limits_{x \to 0} \dfrac{\cos(5x)}{x}$

7. $\lim\limits_{x \to 9^-} \dfrac{|x-9|}{2x-18}$

Table 2.6

x	0.9	0.99	0.999	0.9999
$\dfrac{x^3-1}{|x-1|}$	-2.71	-2.9701	-2.997001	-2.99970001

8. From Table 2.6, given that a limit exists, what is a reasonable expectation for $\lim\limits_{x \to 1^-} \dfrac{x^3-1}{|x-1|}$?

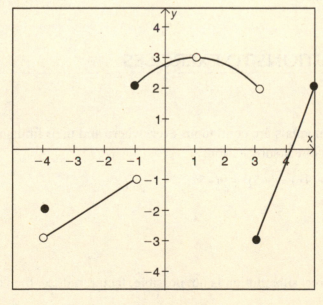

Figure 2.10

Use the graph of $p(x)$ in Figure 2.10 to approximate the following limits. If a limit does not exist, write "DNE."

9. $\lim\limits_{x \to 1} p(x)$

10. $\lim\limits_{x \to -1^+} p(x)$

11. $\lim\limits_{x \to -1} p(x)$

12. $\lim\limits_{x \to 0} p(x)$

13. $\lim\limits_{x \to -4^+} p(x)$

Use the graph of $p(x)$ in Figure 2.10 to answer Exercises 14 and 15.

14. Briefly explain why $\lim\limits_{x \to 3} p(x)$ does not exist.

15. For the graph of $p(x)$ above, list all intervals of continuity.

16. Assume the velocity of a vehicle, as it accelerates, is a continuous function. Explain why a vehicle cannot go from 30 miles per hour to 50 miles per hour without at least once attaining speed of 45 miles per hour during that interval.

2.6 SOLUTIONS TO EXERCISES

1. 70

   All polynomials are continuous everywhere and their limits may be evaluated by direct substitution.

   $$\lim_{x \to -5}(x^2 - 9x) = (-5)^2 - 9(-5)$$
   $$= 70$$

2. 18

   Since direct substitution is not possible, factor, reduce, then evaluate.

$$\lim_{x \to 3} \frac{(x^3 - 9x)}{x - 3} = \lim_{x \to 3} \frac{x(x-3)(x+3)}{x-3}$$

$$= \lim_{x \to 3} \frac{x(x+3)}{1}$$

$$= 3(3+3)$$

$$= 18$$

3. $\dfrac{5}{8}$

Use $\lim\limits_{x \to 0} \dfrac{\sin(a \cdot x)}{b \cdot x} = \dfrac{a}{b}$

$$\lim_{x \to 0} \frac{\sin(5x)}{8x} = \frac{5}{8}$$

4. ∞

There is no analytic approach. Take a numerical approach by testing x values just larger than 7 since the limit is approaching 7 from the right.

$$x = 7.1 \Rightarrow \frac{2}{0.1} = 20$$

$$x = 7.01 \Rightarrow \frac{2}{0.01} = 200$$

$$x = 7.001 \Rightarrow \frac{2}{0.001} = 2000$$

The pattern indicates $\lim\limits_{x \to 7^+} \dfrac{2}{x - 7} = \infty$.

5. 1

If c is a constant, and $f(x)$ is a function with a defined limit, $\lim\limits_{x \to \infty} c^{(f(x))} = c^{\left(\lim\limits_{x \to \infty} f(x)\right)}$.

$$\lim_{x \to \infty} 3^{\left(\frac{1}{x}\right)} = 3^{\left(\lim\limits_{x \to \infty} \frac{1}{x}\right)}$$

$$= 3^0$$

$$= 1$$

6. DNE (does not exist)

$\lim\limits_{x \to 0^+} \dfrac{\cos(5x)}{x} = \infty$ since the ratio approaches 1 divided by an infinitely small positive number. Also, $\lim\limits_{x \to 0^-} \dfrac{\cos(5x)}{x} = -\infty$ since the ratio approaches 1 divided by an infinitely small negative number.

Because the left- and right-hand limits are unequal, $\lim\limits_{x \to 0} \dfrac{\cos(5x)}{x}$ does not exist.

7. $\dfrac{-1}{2}$

For values just less than 9, $x - 9 < 0$; therefore, $|x - 9| = -(x - 9)$.

$$\lim\limits_{x \to 9^-} \frac{|x-9|}{2x-18} = \lim\limits_{x \to 9^-} \frac{-(x-9)}{2(x-9)}$$
$$= \frac{-1}{2}$$

8. -3

Given that a limit exists, a reasonable expectation is $\lim\limits_{x \to 1^-} \dfrac{x^3-1}{|x-1|} = -3$. This can be confirmed analytically.

For values of x just less than 1, $x - 1 < 0$; therefore, $|x - 1| = -(x - 1)$.

$$\lim\limits_{x \to 1^-} \frac{x^3-1}{|x-1|} = \lim\limits_{x \to 1^-} \frac{(x-1)(x^2+x+1)}{-(x-1)}$$
$$= \lim\limits_{x \to 1^-} \frac{x^2+x+1}{-1}$$
$$= -3$$

9. 3

$$\lim\limits_{x \to 1} p(x) = 3$$

The point at (1, 3) does not need to exist for the limit to exist.

10. 2

   $$\lim_{x \to -1^+} p(x) = 2$$

   The function values of p approach 2 for x-values just to the right of $x = -1$.

11. DNE

   $\lim_{x \to -1} p(x)$ does not exist because the limits from the left and right are not equal.

   $$\lim_{x \to -1^+} p(x) = 2 \text{ and } \lim_{x \to -1^-} p(x) = -1$$

12. ≈ 2.75

   $$\lim_{x \to 0} p(x) \approx 2.75$$

   p is continuous at $x = 0$, and $p(0) \approx 2.75$.

13. -3

   $$\lim_{x \to -4^+} p(x) = -3$$

   $p(x)$ approaches -3 as x approaches -4 from the right. The point at $(-4, -2)$ is *not* considered when finding the limit.

14. DNE

   $\lim_{x \to 3} p(x)$ does not exist because the left- and right-hand limits at $x = 3$ are unequal.

15. The intervals of continuity are $(-4, -1) \cup (-1, 1) \cup (1, 3) \cup (3, 5]$.

16. For a continuous function on a closed interval, the Intermediate Value Theorem guarantees that the function will take on all values between the function values of the endpoints at least once. In this case, $v(a) = 30$ mph, and $v(b) = 50$ mph, and all speeds between 30 and 50 mph will be attained in the time interval (a, b) at least once, including 45 mph.

CHAPTER 3

Concepts of the Derivative

CONCEPTS OF THE DERIVATIVE

3.1 INTRODUCTION

With the establishment of limits and continuity, a foundation has been laid for studying the rest of calculus. Calculus is the study of change—both large and infinitesimally small change. In previous courses, you studied the slope of linear functions, but the use of limits in calculus enables mathematicians to study curves with constantly changing slopes. With calculus, one can also apply varying slope to position, velocity, acceleration, and anything else that is in a state of change.

3.2 RATES OF CHANGE

AVERAGE RATE OF CHANGE

Suppose 120 miles is covered in 3 hours on a car trip. It is common to speak of the average speed for the trip as 40 miles per hour. Without question, during the trip the vehicle traveled at speeds other than 40 miles per hour, but based on just the distance covered and the time it took to cover that distance, regardless of what happened during the trip, an average rate can be determined. Anything that changes over time can have an average rate calculated. For example, if the outside temperature increases 12 degrees in 6 hours, the average rate of increase is 2 degrees per hour.

Average Rate of Change

If a quantity Q changes as a function of t on the interval $[t_1, t_2]$, then the average rate of change of Q with respect to t is $\dfrac{\Delta Q}{\Delta t} = \dfrac{Q(t_2) - Q(t_1)}{t_2 - t_1}$.

The formula for the average rate of change should look familiar. It is simply the slope between two points of a function. In previous courses, the majority of functions were of the form $y = f(x)$, and the slope was calculated by using $\dfrac{\Delta y}{\Delta x} = \dfrac{y_2 - y_1}{x_2 - x_1}$. It is important to remember that determining the average rate of change takes no calculus, and that it always measures change over an interval.

EXAMPLE 3.1

Find the average rate of change of $f(x) = 9 - x^2$ on the interval $[-2, 1]$.

SOLUTION

$$\frac{\Delta f}{\Delta x} = \frac{f(1) - f(-2)}{1 - (-2)}$$
$$= \frac{8 - 5}{3}$$
$$= 1$$

The average rate of change of a function must be understood from a graphical standpoint as well. Recall from geometry that the name for a line passing through a circle at two points is a secant. On nonlinear functions, the average rate of change is the slope of the secant connecting any two points on the function. Figure 3.1, the graph of the function in Example 3.1, shows this principle.

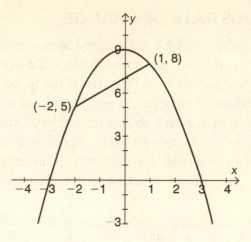

Figure 3.1

Be aware that average rate of change may be called by other names, such as average acceleration. Since acceleration is a change in velocity over time, average acceleration is simply the slope of a secant line between two points on the graph of a velocity function.

EXAMPLE 3.2

For $t \geq 0$, the velocity, in feet per second, of a moving object is $v(t) = t \cdot \sin(t)$. Find the average acceleration over the time interval $t = 0$ to $t = \dfrac{\pi}{2}$ seconds.

SOLUTION

Average acceleration is the slope between points on the velocity function.

$$\frac{v\left(\frac{\pi}{2}\right) - v(0)}{\frac{\pi}{2} - 0} = \frac{\frac{\pi}{2}\sin\left(\frac{\pi}{2}\right) - 0\sin(0) \text{ ft/sec}}{\frac{\pi}{2} \text{ sec}}$$

$$= 1 \text{ ft/sec}^2$$

(Notice the units are units of acceleration!)

INSTANTANEOUS RATE OF CHANGE

During a car trip, when we look at the speedometer, it shows our speed at a moment in time. It is still a rate, just not an average rate of change. It is the rate of change of position at an instant in time. But if the speedometer were broken and we wanted to estimate our speed at any instant, we would have to calculate change in position over a very short time interval. Even though this is still an average rate, over a short enough interval, speed will vary less than over a longer interval and will be a reasonable estimate of instantaneous velocity. Fortunately, the use of limits allows calculation of an instantaneous rate of change for many functions.

Just as before, let Q be a quantity changing over time. In this situation, rather than using t_1 and t_2 as the two moments in time, we will keep track of the change in time by using Δt. The interval over which the quantity is measured will then be from t to $t + \Delta t$.

Instantaneous Rate of Change

If a quantity Q changes as a function of t on the interval $[t, t + \Delta t]$, then the instantaneous rate of change of Q with respect to t is $\lim_{\Delta t \to 0} \dfrac{\Delta Q}{\Delta t} = \dfrac{Q(t + \Delta t) - Q(t)}{\Delta t}$.

With a known function, any of the various methods of evaluating limits may be used to determine an instantaneous rate of change. Without a function, but with numerical data, the instantaneous rate of change at a single moment may be estimated by using the smallest available interval containing the point at which the instantaneous rate is desired.

EXAMPLE 3.3

Find the instantaneous rate of change of $f(x) = x^2 - 3$ when $x = 1$.

SOLUTION

$$\lim_{\Delta x \to 0} \frac{\Delta f}{\Delta x} = \lim_{\Delta x \to 0} \frac{f(1+\Delta x) - f(1)}{(1+\Delta x) - 1}$$

$$= \lim_{\Delta x \to 0} \frac{[(1+\Delta x)^2 - 3] - [1^2 - 3]}{\Delta x}$$

$$= \lim_{\Delta x \to 0} \frac{1 + 2\Delta x + (\Delta x)^2 - 3 - 1 + 3}{\Delta x}$$

$$= \lim_{\Delta x \to 0} \frac{2\Delta x + (\Delta x)^2}{\Delta x}$$

$$= \lim_{\Delta x \to 0} \frac{\Delta x(2 + \Delta x)}{\Delta x}$$

$$= \lim_{\Delta x \to 0} (2 + \Delta x)$$

$$= 2$$

EXAMPLE 3.4

Ignoring air resistance, the number of feet an object will fall in t seconds near the surface of Earth is determined by the relationship $d(t) = 16t^2$. How fast is an object falling 2 seconds after it is dropped?

SOLUTION

Calculate the instantaneous rate of change of position when $t = 2$ seconds.

$$\lim_{\Delta t \to 0} \frac{\Delta d}{\Delta t} = \lim_{\Delta t \to 0} \frac{d(2+\Delta t) - d(2)}{(2+\Delta t) - 2}$$

$$= \lim_{\Delta t \to 0} \frac{[16(2+\Delta t)^2] - [16 \cdot 2^2]}{\Delta t}$$

$$= \lim_{\Delta t \to 0} \frac{64 + 64\Delta t + 16(\Delta t)^2 - 64}{\Delta t}$$

$$= \lim_{\Delta t \to 0} \frac{64\Delta t + 16(\Delta t)^2}{\Delta t}$$

$$= \lim_{\Delta t \to 0} \frac{16\Delta t(4 + \Delta t)}{\Delta t}$$

$$= \lim_{\Delta t \to 0} [16(4 + \Delta t)]$$

$$= 64 \text{ ft/sec}$$

Since distance is measured in feet, and time in seconds, at $t = 2$ seconds, the object is falling at 64 feet per second. Notice again, based on the units, it is reasonable to conclude that the instantaneous rate of change of position at any moment equals velocity at that moment. This is instantaneous velocity, not average velocity.

EXAMPLE 3.5

The speed of a car, measured every 2 seconds for 6 seconds, is shown in Table 3.1.

Table 3.1

Time (sec)	0	2	4	6
Speed (ft/sec)	40	48	60	75

Use the data to estimate the instantaneous rate of change of speed at $t = 3$ seconds.

SOLUTION

Use the smallest interval containing $t = 3$ seconds and calculate the average rate of change of speed during that interval.

$$\frac{(60 - 48) \text{ ft/sec}}{(4 - 2) \text{ sec}} = 6 \text{ ft/sec}^2$$

Remember that the problem asked for an estimate, and even though a limit cannot be taken, the time interval used is the smallest available around 3 seconds. Also notice the units. The rate of change of velocity is acceleration.

3.3   INTRODUCTION TO DERIVATIVES

SLOPE AT A POINT

As the interval over which an average rate of change is calculated becomes shorter and shorter, the secant line begins to resemble a tangent line to the curve. If the sequence of the slopes of the secants has a limit, that limit is considered the slope of the function at a single point and is called the **derivative** of the function at that point.

Derivative of a Function at Point $x = a$

The derivative of a function f at a given point $x = a$, where a is a constant, is $f'(a) = \lim\limits_{h \to 0} \dfrac{f(a+h)-f(a)}{h}$. $f'(a)$ is read "f prime of a."

It is important to understand the geometric significance of the limit given in the definition of the derivative. Without the limit, the difference quotient is the average rate of change of the function over an interval, and its geometric meaning is the slope of a secant line connecting two points on the graph of the function. With the limit, it is the instantaneous rate of change of the function at a single point $(a, f(a))$, and its geometric meaning is the slope of the tangent line to the function at the point where $x = a$. Figure 3.2 shows the stages of letting h approach 0, and the progression of the secant line "becoming" the tangent line. As h, the horizontal difference between points P and Q, approaches 0, point Q moves toward point P, and the limit of the sequence of slopes of those lines becomes the slope of the tangent line to the graph of $f(x)$ at point P.

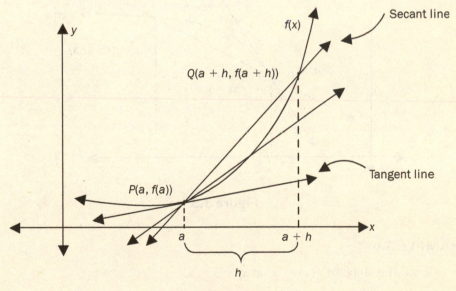

Figure 3.2

An alternate form of the definition of the slope of a function at a given point is based on the same geometric principle. It is sometimes referred to as the "x approaches a form." A good calculus foundation requires understanding both definitions.

Derivative of a Function at Point $x = a$ (x approaches a form)

The derivative of a function f at a given point $x = a$, where a is a constant, is $f'(a) = \lim\limits_{x \to a} \dfrac{f(x) - f(a)}{x - a}$.

In this definition, a is still a known value of the domain at which the slope of the curve is desired, and $(x, f(x))$ is any general point on the graph. The slope of the secant between $(a, f(a))$ and $(x, f(x))$ is $\dfrac{\Delta y}{\Delta x} = \dfrac{f(x) - f(a)}{x - a}$. The limit then causes point $(x, f(x))$ to move closer and closer to $(a, f(a))$ and the limit of the slopes of the secants once again becomes the slope of the tangent. Although there are some small changes from the last figure, Figure 3.3 should look very familiar! Imagine point Q sliding along the graph of $f(x)$ toward point P.

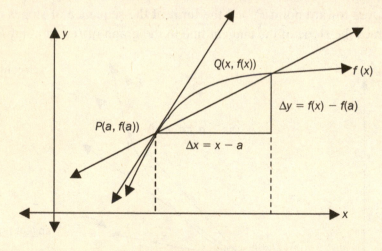

Figure 3.3

EXAMPLE 3.6

Find the slope of $f(x) = x^2$ at $x = 3$.

SOLUTION

The known value is $a = 3$. Use $f'(a) \lim\limits_{h \to 0} \dfrac{f(a+h) - f(a)}{h}$.

$$f'(3) = \lim_{h \to 0} \frac{f(3+h) - f(3)}{h}$$

$$= \lim_{h \to 0} \frac{f(3+h)^2 - (3)^2}{h}$$

$$= \lim_{h \to 0} \frac{9 + 6h + h^2 - 9}{h}$$

$$= \lim_{h \to 0} \frac{h(6+h)}{h}$$

$$= \lim_{h \to 0} (6+h)$$

$$= 6$$

EXAMPLE 3.7

Use the alternate definition of a derivative to find the slope of $f(x) = x^2$ at $x = 3$.

SOLUTION

Use $f'(a) = \lim\limits_{x \to a} \dfrac{f(x) - f(a)}{x - a}$ with $a = 3$.

$$f'(3) = \lim_{x \to 3} \frac{f(x) - f(3)}{x - 3}$$

$$= \lim_{x \to 3} \frac{x^2 - 9}{x - 3}$$

$$= \lim_{x \to 3} \frac{(x-3)(x+3)}{x - 3}$$

$$= \lim_{x \to 3} (x + 3)$$

$$= 6$$

An important skill for success in calculus is to develop familiarity with the vocabulary. This will help to clarify the wide variety of ways to represent or state similar ideas. For instance, in each of the previous two examples, the result shows that the slope of a line tangent to the graph of $f(x) = x^2$ at $(3, 9)$

is 6. Alternately stated, the slope of $f(x)$ at $x = 3$ is 6. One might also just write $f'(3) = 6$. Finally, with a focus on the slope representing the change in y, divided by the change in x, it can also be said that the instantaneous rate of change of the y-coordinate with respect to the x-coordinate is 6, or $\dfrac{dy}{dx} = 6$.

THE DERIVATIVE AS A FUNCTION

If the slope of a function can be calculated at a single value of x, the next logical step is to generalize this to the entire function. Essentially, this means calculating the slope at all the points of the domain. Naturally, doing so would be impossible since the domains of many functions have an infinite number of elements. But this can still be accomplished by simply generalizing the definition of the derivative at a known point $x = a$. Example 3.6 is worked out here by leaving a as a variable instead of using $a = 3$.

$$\begin{aligned}
f'(a) &= \lim_{h \to 0} \frac{f(a+h) - f(a)}{h} \\
&= \lim_{h \to 0} \frac{(a+h)^2 - (a)^2}{h} \\
&= \lim_{h \to 0} \frac{a^2 + 2ah + h^2 - a^2}{h} \\
&= \lim_{h \to 0} \frac{h(2a + h)}{h} \\
&= \lim_{h \to 0} (2a + h) \\
&= 2a
\end{aligned}$$

This result means that at any point on the function $f(x) = x^2$, the slope of the line tangent to the function is twice the x-coordinate. The derivative of $f(x) = x^2$ is $2x$, or $f'(x) = 2x$, or $\dfrac{dy}{dx} = 2x$.

Derivative of a Function

The derivative of a function f is $f'(x) = \lim\limits_{h \to 0} \dfrac{f(x+h) - f(x)}{h}$ for all points of the domain of f where the limit exists.

EXAMPLE 3.8

For $g(x) = \dfrac{1}{x}$ on $x > 0$, find $g'(x)$.

SOLUTION

$$g'(x) = \lim_{h \to 0} \frac{g(x+h) - g(x)}{h}$$

$$= \lim_{h \to 0} \frac{\dfrac{1}{x+h} - \dfrac{1}{x}}{h}$$

$$= \lim_{h \to 0} \frac{\dfrac{1}{x+h} - \dfrac{1}{x}}{h} \cdot \frac{x(x+h)}{x(x+h)}$$

$$= \lim_{h \to 0} \frac{x - (x+h)}{hx(x+h)}$$

$$= \lim_{h \to 0} \frac{-h}{hx(x+h)}$$

$$= \lim_{h \to 0} \frac{-1}{x(x+h)}$$

$$= \frac{-1}{x^2}$$

Therefore, the slope of $g(x) = \dfrac{1}{x}$ on the given domain is the opposite reciprocal of the square of the x-coordinate. For example, $g'(2) = \dfrac{-1}{4}$. The graph of $g(x) = \dfrac{1}{x}$ and the line tangent at $x = 2$ are shown in Figure 3.4. Notice that the x- and y-intercepts of the line show the slope to be $\dfrac{-1}{4}$.

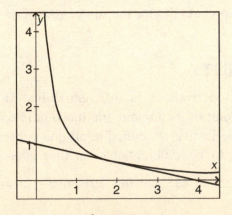

Figure 3.4

EXAMPLE 3.9

A particle is moving along the x-axis. Its position as a function of time is defined as $p(t) = \sqrt{t}$. Find an expression for the instantaneous rate of change of position at any moment $t > 0$.

SOLUTION

This is just a change of variables. Use $p'(t) = \lim\limits_{h \to 0} \dfrac{p(t+h) - p(t)}{h}$

$$p'(t) = \lim\limits_{h \to 0} \frac{\sqrt{t+h} - \sqrt{t}}{h}$$

Simplifying the limit above is done by a special process called "rationalizing the numerator" by multiplying the numerator and denominator by the conjugate of the numerator, $\sqrt{t+h} + \sqrt{t}$.

$$p'(t) = \lim\limits_{h \to 0} \frac{\sqrt{t+h} - \sqrt{t}}{h} \cdot \frac{\sqrt{t+h} + \sqrt{t}}{\sqrt{t+h} + \sqrt{t}}$$

$$= \lim\limits_{h \to 0} \frac{t+h-t}{h(\sqrt{t+h} + \sqrt{t})}$$

$$= \lim\limits_{h \to 0} \frac{h}{h(\sqrt{t+h} + \sqrt{t})}$$

$$= \lim\limits_{h \to 0} \frac{1}{\sqrt{t+h} + \sqrt{t}}$$

$$= \frac{1}{2\sqrt{t}}$$

This means the instantaneous rate of change of the position of the particle at any time, t, is the reciprocal of twice the square root of t.

DIFFERENTIABILITY

A term related to derivatives is differentiability. If the derivative of a function exists at a point of its domain, the function is said to be *differentiable* at that point. If the derivative is defined at all points of a given interval, then the function is said to be differentiable on that interval. In Example 3.9, $p'(t) = \dfrac{1}{2\sqrt{t}}$. It is clear to see that the derivative is defined only for positive values of t. Therefore $p(t)$ is differentiable for $t > 0$; it is not differentiable for $t \leqslant 0$.

Remember that the derivative is defined by using a limit. If for any reason that limit fails to exist, then the derivative does not exist.

Table 3.2 summarizes common reasons for the derivative to fail to exist at a given point $x = a$

Table 3.2

Reason and sketch of $f(x)$ at $x = a$	Explanation
Corner	The limit of the slopes of the secants from the left and right sides of a are different. $$\lim_{h \to 0^-} = \frac{f(a+h) - f(a)}{h} \neq$$ $$\lim_{h \to 0^+} = \frac{f(a+h) - f(a)}{h}$$
Cusp	The limits of the slopes of the secants from the left and right are opposite. $$\lim_{h \to 0^-} = \frac{f(a+h) - f(a)}{h} =$$ $$-\lim_{h \to 0^+} = \frac{f(a+h) - f(a)}{h}$$
Discontinuity	$f(a)$ does not exist, so the limit cannot be evaluated. The discontinuity can be removable or nonremovable.
Vertical Tangent	f is continuous at a, but the limit of the slopes of the secants is infinite.

The discontinuity case in if-then form reads, "If a function is not continuous at a point, then it is not differentiable." The contrapositive of this statement is logically true, and is an important concept: "If a function is differentiable at a point, then it is continuous." Another way to summarize this relationship between continuity and differentiability is to say that differentiability at a point guarantees continuity, but continuity does not guarantee differentiability.

ENDPOINTS

In calculus, endpoints of intervals, whether open or closed, tend to require special treatment. This is certainly true when talking about differentiability. On any closed interval $[a, b]$, in order to define the derivative at either endpoint, it must be acknowledged that the limit used to define the derivative cannot be examined from both the left and right sides. Accordingly, the existence of a one-sided limit at the endpoint is sufficient to confirm differentiability.

If a is the left endpoint of the interval, $f'(a) = \lim\limits_{h \to 0^+} \dfrac{f(a+h) - f(a)}{h}$.

If b is the right endpoint of the interval, $f'(b) = \lim\limits_{h \to 0^-} \dfrac{f(b+h) - f(b)}{h}$.

The alternate form of the definition of the derivative looks similar to the one above.

If a is the left endpoint of the interval, $f'(a) = \lim\limits_{h \to a^+} \dfrac{f(x) - f(a)}{x - a}$.

If b is the right endpoint of the interval, $f'(b) = \lim\limits_{x \to b^-} \dfrac{f(x) - f(b)}{x - b}$.

One-sided limits are also used to determine whether a piecewise-defined function is differentiable at key points.

EXAMPLE 3.10

Let $g(x) = \begin{cases} 3x-1 & x < 1 \\ x^2+1 & x \geq 1 \end{cases}$

Determine whether g is differentiable at $x = 1$.

SOLUTION

The first step is to determine continuity at $x = 1$. If this fails, there is no differentiability.

$$\lim_{x \to 1^-}(3x-1) = 2$$

$$\lim_{x \to 1^+}(x^2+1) = 2$$

Since the limits are equal, $g(x)$ is continuous at $x = 1$.

The second step is to determine whether the derivatives are the same from each side.

$$\lim_{h \to 0^-} \frac{[3(x+h)-1]-[3x-1]}{h} = \lim_{h \to 0^-} \frac{3x+3h-1-3x+1}{h}$$

$$= \lim_{h \to 0^-} \frac{3h}{h}$$

$$= 3$$

(This should be no surprise, since the left portion of g is linear with a slope of 3.)

$$\lim_{h \to 0^+} \frac{[(1+h)^2+1]-[1^2+1]}{h} = \lim_{h \to 0^+} \frac{1+2h+h^2+1-2}{h}$$

$$= \lim_{h \to 0^+} \frac{h(2+h)}{h}$$

$$= \lim_{h \to 0^+}(2+h)$$

$$= 2$$

Since the limit of the slopes of the tangents as x approaches 1 from the left and right are not equal, $g(x)$ is not differentiable at $x = 1$.

3.4 LINEAR APPROXIMATIONS

One of the very interesting results of being able to think about functions "microscopically" by examining slopes over extremely small intervals is that graphs that have always naturally appeared curved can now been seen or thought of as straight. A term commonly applied to this is "local linearity." If a function is differentiable on a given domain, it will have local linearity. One application of this is that functions, whether simple or very complicated, can be approximated by using straight lines.

The graphs shown in Figures 3.5 and 3.6 are two different views of $g(x) = x^2$. Point (1, 1) has been marked on each graph. The main difference between the two is the viewing window. Figure 3.5 has a relatively normal viewing window of five units in each direction from the origin. Figure 3.6 has "zoomed in" on the point (1, 1) and shows the function on a domain from about [0.5, 1.5]. Even though this is not exactly a microscopic look, one should see the linear nature of the graph already beginning to take shape.

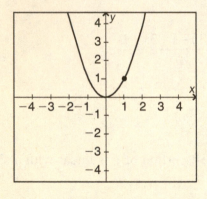

Figure 3.5

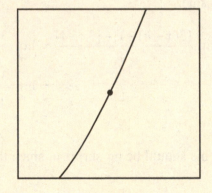

Figure 3.6

Conversely, if a function is not differentiable at a certain point, it will never approach linearity there. This same graphical comparison is now used on $h(x) = 1 + |x|$ in Figures 3.7 and 3.8. In this case, however, no matter how closely one looks at the point (0, 1), the curve will never appear straight.

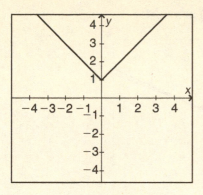

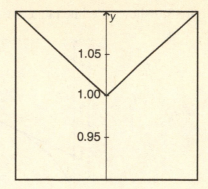

Figure 3.7   **Figure 3.8**

In Section 3.3, the derivative is defined as the slope of the line tangent to a graph at a given point. The tangent line at any given point is the linear approximation of the function over relatively small intervals. For nonlinear functions, the approximation is accurate only over small intervals because any nonlinear function has a varying slope, whereas a line has a constant slope.

EXAMPLE 3.11

Find the line tangent to $g(x) = \sqrt{x}$ at $x = 4$.

SOLUTION

An equation of a line can be written by using the slope of the line and any point on the line. The function provides the point, and the derivative of the function provides the slope.

$g(4) = 2$

The point of tangency to $g(x)$ is $(4, 2)$.

Applying the result of Example 3.9 to g, $g'(x) = \dfrac{1}{2\sqrt{x}}$, so $g'(4) = \dfrac{1}{4}$.

Thus the slope of the tangent at $x = 4$ is $\dfrac{1}{4}$.

Using the point-slope form, the tangent line equation is $y - 2 = \dfrac{1}{4}(x - 4)$.

Figure 3.9 shows the line tangent to $g(x)$ at $x = 4$.

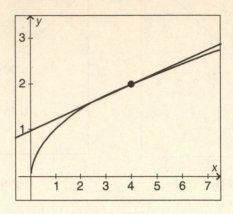

Figure 3.9

Figure 3.9 reinforces the idea that near the point of tangency, the line provides reasonably good approximations of the function, but as the domain around $x = 4$ widens, the graphs diverge.

3.5 EXERCISES

Work the following exercises without a calculator. Solutions follow this section.

1. In a sentence or two, describe the difference between the average rate of change of a function and the instantaneous rate of change of a function.

2. Find the average rate of change of $h(x) = x^3 - 2x$ on the domain $[1, 3]$.

3. A circular metal plate was heated. When the temperature at the center reached 100°C, the heat source was removed. Table 3.3 shows the temperature at the center of the plate as it cooled over time. Use the table to approximate the instantaneous rate of cooling of the plate at time $t = 5$ minutes.

Table 3.3

Time (min)	0	2	4	6	8	10	12
Temperature (˚C)	100	74	58	46	38	32	28

4. Use Table 3.3 to determine the average rate of cooling from time 2 minutes to 12 minutes and write a brief explanation to interpret the meaning of the result.

5. Find the instantaneous rate of change of the function $f(x) = 2x^2 + x$ at $x = 1$.

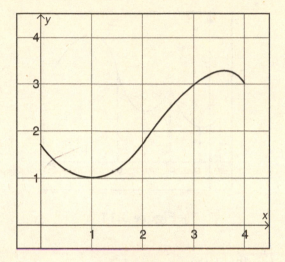

Figure 3.10

6. On the graph shown in Figure 3.10, put each of the following letters at the place or places that best meet the given condition.

    A: The slope of the graph is 0.

    B: The instantaneous rate of change of the function is greatest.

    C: The instantaneous rate of change equals the average rate of change for the entire domain [0, 4].

    D: The instantaneous rate of change is most negative.

7. If $k(x) = \dfrac{3}{x+2}$, find $k'(x)$.

8. For the time interval [0, 4] seconds, the velocity of a cyclist, in feet per second, is

    $$v(t) = \begin{cases} 0.5(t+1)^2, & 0 \le t \le 2 \\ 3t - 1.5, & 2 < t \le 4 \end{cases}$$

    Find the absolute value of the difference in the cyclist's acceleration at the beginning and at the end of the given interval.

9. For the graph in Figure 3.11, list the letter for each place where the function is *not* differentiable.

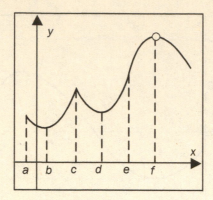

Figure 3.11

10. If $f(x) = \sqrt{\dfrac{x}{2}}$, find $f'(8)$.

11. The graph of $b'(x)$ is shown in Figure 3.12. If $b(2) = -7$, find the equation of the line tangent to b at $x = 2$.

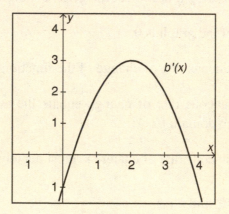

Figure 3.12

3.6 SOLUTIONS TO EXERCISES

1. The average rate of change measures the change in a function over an interval. The result is the slope of the secant joining the points at the endpoints of the interval. An instantaneous rate of change measures the rate of change of a function at a specific point on the function. The result is the slope of the line tangent to the graph of the function at the given point.

2. 11

Average rate of change $= \dfrac{h(3) - h(1)}{3 - 1}$; $h(x) = x^3 - 2x$

$$\dfrac{h(3) - h(1)}{3 - 1} = \dfrac{(3^3 - 2 \cdot 3) - (1^3 - 2 \cdot 1)}{2}$$

$$= \dfrac{21 - (-1)}{2}$$

$$= 11$$

3. –6 degrees per minute

The best estimate is found by calculating the slope over the smallest interval containing $t = 5$ minutes. In this situation, that is between 4 and 6 minutes.

$$\dfrac{46 - 58}{6 - 4} = -6 \text{ degrees per minute.}$$

4. –4.6 degrees per minute

Using the values in Table 3.3, at 2 minutes the temperature is 74°C, and at 12 minutes the temperature is 28°C.

The average rate of change is the change in temperature over the change in time.

$$\dfrac{28 - 74}{12 - 2} = -4.6 \text{ degrees per minute}$$

During the time interval from 2 to 12 minutes, the temperature of the center of the plate decreased an average of 4.6 degrees per minute.

5. 5

The instantaneous rate of change of the function $f(x) = 2x^2 + x$ at $x = 1$ is the derivative at that point.

$$f'(1) = \lim_{x \to 1} \dfrac{f(x) - f(1)}{x - 1}$$

$$= \lim_{x \to 1} \dfrac{2x^2 + x - 3}{x - 1}$$

$$\lim_{x \to 1} \dfrac{(2x + 3)(x - 1)}{x - 1}$$

$$\lim_{x \to 1} (2x + 3)$$

$$= 5$$

6.

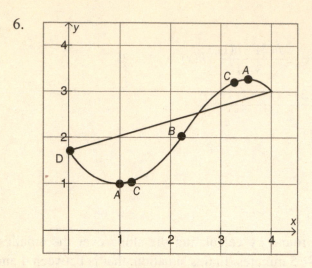

Figure 3.13

This problem is all about envisioning tangent lines to the curve since the slope of the tangent line is the instantaneous rate of change of the function.

For A, consider where tangent lines are horizontal, so the slope is 0.

For B, look for where the slope of a tangent line is most positive.

For C, the secant line has been added to the drawing because its slope is the average rate of change. Points C have been placed on the curve at the two locations where the slope of the tangent line would equal the slope of the secant line.

For D, look for where the slope of the tangent is most negative.

7. $k'(x) = \dfrac{-3}{(x+2)^2}$

If $k(x) = \dfrac{3}{x+2}$, then

$$k'(x) = \lim_{h \to 0} \frac{k(x+h) - k(x)}{h}$$

$$= \lim_{h \to 0} \frac{\frac{3}{x+h+2} - \frac{3}{x+2}}{h} \cdot \frac{(x+h+2)(x+2)}{(x+h+2)(x+2)}$$

$$= \lim_{h \to 0} \frac{3(x+2) - 3(x+h+2)}{h(x+h+2)(x+2)}$$

$$= \lim_{x \to 0} \frac{3x + 6 - 3x - 3h - 6}{h(x+h+2)(x+2)}$$

$$= \lim_{h \to 0} \frac{-3h}{h(x+h+2)(x+2)}$$

$$= \lim_{h \to 0} \frac{-3}{(x+h+2)(x+2)}$$

$$= \frac{-3}{(x+2)^2}$$

8. 2 ft/sec²

$$v(t) = \begin{cases} 0.5(t+1)^2, & 0 \le t \le 2 \\ 3t - 1.5, & 2 < t \le 4 \end{cases}$$

Acceleration is the instantaneous rate of change of velocity with respect to time. To answer the problem, calculate $|v'(4) - v'(0)|$. Finding $v'(t)$ at each value will require using the appropriate portion of the piecewise function. $v'(4) = 3$, since $v(t) = 3t - 1.5$ is linear, so it has a constant slope of 3.

$$v'(0) = \lim_{t \to 0^+} \frac{v(t) - v(0)}{t - 0}$$

$$= \lim_{t \to 0^+} \frac{0.5(t+1)^2 - 0.5}{t}$$

$$= \lim_{t \to 0^+} \frac{0.5(t^2 + 2t + 1) - 0.5}{t}$$

$$= \lim_{t \to 0^+} \frac{0.5t^2 + t + 0.5 - 0.5}{t}$$

$$= \lim_{t \to 0^+} \frac{t(0.5t + 1)}{t}$$

$$= \lim_{t \to 0^+} t(0.5t + 1)$$

$$= 1$$

$$|v'(4) - v'(0)| = |3 - 1| = 2 \text{ ft/sec}^2$$

9. The function is *not* differentiable at points *c* and *f*.

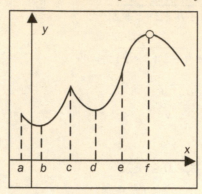

Figure 3.14

c is a corner, so the limits of the slopes of the tangents from the left and the right side of *c* will never be the same.

f is a point of discontinuity. A function must be continuous at a given point to have a derivative at that point.

The function *is* differentiable at *a*, since it is an endpoint, and only the one-sided limit must exist.

10. $\dfrac{1}{8}$

Think of $f(x) = \sqrt{\dfrac{x}{2}}$ as $f(x) = \dfrac{1}{\sqrt{2}} \cdot \sqrt{x}$.

$$f'(8) = \lim_{x \to 8} \frac{f(x) - f(8)}{x - 8}$$

$$= \lim_{x \to 8} \frac{\frac{1}{\sqrt{2}}\sqrt{x} - \frac{1}{\sqrt{2}}\sqrt{8}}{x - 8}$$

$$= \frac{1}{\sqrt{2}} \lim_{x \to 8} \frac{\sqrt{x} - \sqrt{8}}{x - 8} \cdot \frac{\sqrt{x} + \sqrt{8}}{\sqrt{x} + \sqrt{8}}$$

$$= \frac{1}{\sqrt{2}} \lim_{x \to 8} \frac{(x - 8)}{(x - 8)(\sqrt{x} + \sqrt{8})}$$

$$= \frac{1}{\sqrt{2}} \lim_{x \to 8} \frac{1}{(\sqrt{x} + \sqrt{8})}$$

$$= \frac{1}{\sqrt{2}} \cdot \frac{1}{2\sqrt{8}}$$

$$= \frac{1}{2\sqrt{16}}$$

$$= \frac{1}{8}$$

11. $y + 7 = 3(x - 2)$ or $y = 3x - 13$

Since $b(2) = -7$, the only additional information required to write the equation of the tangent line is its slope at $x = 2$. $b'(2)$ is the y-coordinate at $x = 2$ on the given graph, so $b'(2) = 3$.

In point-slope form, the equation of the tangent line is $y + 7 = 3(x - 2)$, or simplifying to slope-intercept form, $y = 3x - 13$.

CHAPTER 4

Rules of Differentiation

RULES OF DIFFERENTIATION

4.1   INTRODUCTION

It would be an extremely tedious process to always determine derivatives by the use of limits. As a result, derivative rules for a wide variety of functions can be established by using limits, and those rules become the more efficient tools for finding derivatives. Since this text is not intended to be a first exposure to calculus, most of the derivative rules will be given without proof, although a few key derivatives will be established through limits.

4.2   DERIVATIVES OF POLYNOMIALS

Some derivatives can be determined just from the knowledge that a derivative represents the slope of the function. For instance, the derivative of a constant is zero, since the slope of the graph of $y = c$, where c is a constant, is 0.

Derivative of a Constant

If c is any constant, and $y = c$, then $\dfrac{dy}{dx} = \dfrac{d}{dx}(c) = 0$.

Using similar reasoning, the derivative of any linear function is the slope of that function.

Derivative of a Linear Function

If m is any constant, and $y = mx$, then $\dfrac{dy}{dx} = \dfrac{d}{dx}(mx) = m$.

The derivative of a power of x is also relatively straightforward, but perhaps not intuitively understood. If m is any natural number, and $f(x) = x^m$ then $\dfrac{df}{dx} = \dfrac{d}{dx}(x^m) = m \cdot x^{(m-1)}$.

The proof of this rule utilizes the definition of the derivative and the binomial expansion pattern from a prerequisite course.

$$
\begin{aligned}
\lim_{h \to 0} \frac{(x+h) - x^m}{h} &= \lim_{h \to 0} \frac{(x^m + mx^{m-1}h + {}_mC_2 x^{m-2}h^2 + \ldots + mx^1 h^{m-1} + h^m) - x^m}{h} \\
&= \lim_{h \to 0} \frac{mx^{m-1}h + {}_mC_2 x^{m-2}h^2 + \ldots + mx^1 h^{m-1} + h^m}{h} \\
&= \lim_{h \to 0} \frac{h(mx^{m-1} + {}_mC_2 x^{m-2}h + \ldots + mx^1 h^{m-2} + h^{m-1})}{h} \\
&= \lim_{h \to 0}(mx^{m-1} + {}_mC_2 x^{m-2}h + \ldots + mx^1 h^{m-2} + h^{m-1}) \\
&= mx^{m-1}
\end{aligned}
$$

Although the binomial expansion from courses prior to calculus applies only to exponents that are natural numbers, we will extend the derivative rule (without proof) to any exponents that are elements of the real numbers.

Derivative of a Power

If m is any real number, and $f(x) = x^m$, then
$\dfrac{df}{dx} = \dfrac{d}{dx}(x^m) = m \cdot x^{(m-1)}$.

The derivative of any constant multiple of a power of x is established by factoring the constant out of the limit. Therefore, the coefficient simply multiplies by the power.

Derivative of a Constant Times a Function

If c and m are constants and $y = cx^m$, then
$$\frac{dy}{dx} = \frac{d}{dx}(c \cdot x^m) = c \cdot m \cdot x^{m-1}.$$

Extending derivatives of monomials to polynomials is also elementary. Since a polynomial is the sum of monomials, and derivatives are proven by limits, when each is defined, the limit of a sum equals the sum of the limits.

Derivative of a Sum

If f and g are functions of x, then $\dfrac{d}{dx}(f + g) = \dfrac{df}{dx} + \dfrac{dg}{dx}$.

Example 4.1 makes use of all the properties of derivatives presented in this section: a sum of a constant, a linear term, and a monomial with a coefficient other than 1.

EXAMPLE 4.1

If $y = 7x^2 + 5x - 4$, find $\dfrac{dy}{dx}$.

SOLUTION

$$\frac{dy}{dx} = \frac{d}{dx}(7x^2) + \frac{d}{dx}(5x) - \frac{d}{dx}(4)$$
$$= 7 \cdot 2x^{2-1} + 5 - 0$$
$$= 14x + 5$$

Frequently, the ease with which a derivative can be found depends upon seeing the problem in the appropriate form. In general, writing radical expressions by using fractional exponents will make it easier to apply the derivative rule.

EXAMPLE 4.2

Find the slope of $p(x) = 6\sqrt{x^3}$ at $x = 4$.

SOLUTION

Rewrite $\sqrt{x^3}$ as $x^{\left(\frac{3}{2}\right)}$.

$$\frac{dp}{dx} = \frac{d}{dx}\left(6x^{\left(\frac{3}{2}\right)}\right)$$

$$= 6 \cdot \frac{3}{2} x^{\left(\frac{3}{2}-1\right)}$$

$$= 9x^{\frac{1}{2}}$$

At $x = 4$, $\frac{dp}{dx} = 9\sqrt{4} = 18$.

EXAMPLE 4.3

If $f(x) = \frac{5}{x^2}$, find $f'(x)$.

SOLUTION

Rewrite $\frac{5}{x^2}$ as $5x^{-2}$.

$$f'(x) = \frac{d}{dx}(5x^{-2})$$

$$= 5 \cdot -2x^{-2-1}$$

$$= -10x^{-3}$$

$$= \frac{-10}{x^{-3}}$$

4.3   DERIVATIVES OF PRODUCTS AND QUOTIENTS

The derivative of a *product of functions* must be approached differently than one might assume. The natural inclination would be to find the product of the derivatives of each factor. A simple counterexample shows the fallacy in that thinking.

EXAMPLE 4.4

Is the derivative of a product of functions equal to the product of the derivatives of each factor?

SOLUTION

Let $h(x) = (3x + 1)(2x + 5)$.

Using flawed reasoning, $\dfrac{d}{dx}(3x+1) = 3$, and $\dfrac{d}{dx}(2x+5) = 2$, so one might think $\dfrac{dh}{dx} = 6$.

Of course, it is easy to expand the product to $h(x) = 6x^2 + 17x + 5$.

Now using previously established rules, $\dfrac{dh}{dx} = 12x + 17$.

Therefore, the derivative of a product of functions does not equal the product of the derivatives of each factor.

Since the proof is rather lengthy, the product rule is presented without proof.

Product Rule

If $f(x)$ and $g(x)$ are differentiable functions of x, and $y = f(x) \cdot g(x)$, then $\dfrac{dy}{dx} = \dfrac{d}{dx}(f(x) \cdot g(x)) = f(x) \cdot \dfrac{dg}{dx} + g(x) \cdot \dfrac{df}{dx}$.

This rule can also be applied to a product of more than two functions. For instance, if f, g, and h are all differentiable functions of x, and $y = f \cdot g \cdot h$, then $\dfrac{dy}{dx} = f \cdot g \cdot \dfrac{dh}{dx} + f \cdot h \cdot \dfrac{dg}{dx} + g \cdot h \cdot \dfrac{df}{dx}$. Essentially, only one function is differentiated at a time, while the other functions "wait their turns!" Now let's revisit the previous example and apply the correct product rule.

EXAMPLE 4.5

Find the derivative of $h(x) = (3x + 1)(2x + 5)$, first by expanding and differentiating term by term, and then by using the product rule.

SOLUTION

$h(x) = 6x^2 + 17x + 5$ so $\dfrac{dh}{dx} = 12x + 17$.

Using the product rule,

$$\frac{d}{dx}[(3x+1)(2x+5)] = (3x+1)\frac{d}{dx}(2x+5) + (2x+5)\frac{d}{dx}(3x+1)$$
$$= (3x+1)\cdot 2 + (2x+5)\cdot 3$$
$$= 6x+2+6x+15$$
$$= 12x+17$$

By either method, $h'(x) = 12x + 17$.

As one might expect, there is also a special formula for differentiating quotients of functions. The derivative is not simply the quotient of the individual derivatives.

Quotient Rule

> Let u and v be differentiable functions of x. If $y = \left(\dfrac{u}{v}\right)$, and
> $v(x) \neq 0$, then $\dfrac{dy}{dx} = \dfrac{d}{dx}\left(\dfrac{u}{v}\right) = \dfrac{v \cdot \frac{du}{dx} - u \cdot \frac{dv}{dx}}{v^2}$.

This is somewhat more complicated than the product rule and will take a bit more effort to learn and memorize. Sometimes associating verbalizations can help in that process. For obvious reasons, name the denominator LOW, and the numerator HIGH. The quotient rule can then be spoken as, "LOW, d HIGH, minus HIGH, d LOW, all over LOW squared." The "d" refers to the derivative of that particular part of the quotient. Again, a well-chosen example can provide a bit of confidence in the user that this formula truly works.

EXAMPLE 4.6

Find the derivative of $y = \dfrac{x^3 + 5x^2}{x}$ first by reducing and differentiating term by term, and then by using the quotient rule.

SOLUTION

$$y = \frac{x^3 + 5x^2}{x} = x^2 + 5x, \text{ so } \frac{dy}{dx} = 2x + 5.$$

By the quotient rule,

$$\frac{d}{dx}\left(\frac{x^3 + 5x^2}{x}\right) = \frac{x \cdot \frac{d}{dx}(x^3 + 5x^2) - (x^3 + 5x^2)\frac{d}{dx}(x)}{x^2}$$

$$= \frac{x(3x^2 + 10x) - (x^3 + 5x^2) \cdot 1}{x^2}$$

$$= \frac{3x^3 + 10x^2 - x^3 - 5x^2}{x^2}$$

$$= \frac{2x^3 + 5x^2}{x^2}$$

$$= 2x + 5$$

Clearly, it was easier here to arrive at the derivative by simplifying the quotient prior to differentiating, but this is not always the case. There are times when the quotient either cannot be simplified, or when it is unreasonably complicated to simplify before differentiating. For instance, if we seek $h'(x)$ for $h(x) = \frac{x^2 - 2x + 8}{x^3 + x}$, attempting to simplify the quotient first would be a very unpleasant task! On the other end of the spectrum, the quotient rule is often used when it is unnecessary. Remember to look for a *quotient of functions*. If either the numerator or denominator of a fraction is a constant, the quotient rule is not necessary.

EXAMPLE 4.7

Find $f'(x)$ if $f(x) = \frac{2}{x^3}$.

SOLUTION

$$f(x) = \frac{2}{x^3} = 2x^{-3}$$

$$\frac{d}{dx}(2x^{-3}) = 2 \cdot -3x^{-4}$$

$$= \frac{-6}{x^4}$$

$$f'(x) = \frac{-6}{x^4}$$

Essentially, knowing when to use the product or quotient rules, or when to take an alternate approach, is a matter of practice and experience. It is good practice to pause and think about each approach before attempting the problem. Ask yourself questions such as, "Will the expression simplify nicely to something that is easy to differentiate?" or "Is the quotient rule (or product rule) really necessary to find the derivative?"

4.4   DERIVATIVES OF TRIGONOMETRIC FUNCTIONS

To this point, all functions used have been relatively simple polynomials or rational expressions. Of course, plenty of other types of functions—trigonometric, logarithmic, exponential, and more—need to be considered in the study of calculus. The proofs of the derivatives of some of those functions are very complex, but proving the derivative of the sine function is a worthy endeavor.

Start with the definition of the derivative,

$$\frac{d}{dx}[\sin(x)] = \lim_{h \to 0} \frac{\sin(x+h) - \sin(x)}{h}.$$

Use the expansion of the sine of a sum.

$$\lim_{h \to 0} \frac{\sin(x+h) - \sin(x)}{h} = \lim_{h \to 0} \frac{\sin(x)\cos(h) + \cos(x)\sin(h) - \sin(x)}{h}$$

Regroup and rewrite.

$$\lim_{h \to 0} \frac{\sin(x)\cos(h) + \cos(x)\sin(h) - \sin(x)}{h}$$

$$= \lim_{h \to 0} \frac{\sin(x)[\cos(h) - 1]}{h} + \lim_{h \to 0} \frac{\cos(x)\sin(h)}{h}$$

$$= \sin(h) \lim_{h \to 0} \frac{[\cos(h) - 1]}{h} + \cos(x) \lim_{h \to 0} \frac{\sin(h)}{h}$$

$$= \sin(x) \cdot 0 + \cos(x) \cdot 1$$

$$= \cos(x)$$

A jump in logic was made by substituting 0 for $\lim\limits_{h\to 0} \dfrac{[\cos(h)-1]}{h}$, but if you are curious, try a brief numerical exploration with a calculator to confirm the limit. Notice also that $\lim\limits_{h\to 0} \dfrac{\sin(h)}{h}$ was replaced with 1. This limit was proved in Chapter 2.

By the definition of the derivative, $\dfrac{d}{dx}[\sin(x)] = \cos(x)$.

By the same process, it could be shown (but will not here) that the derivative of $\cos(x)$ is $-\sin(x)$. Once these two derivatives are established, the derivatives of the other four basic trigonometric functions can be determined by using the quotient rule. One function, the tangent, is shown below.

$$
\begin{aligned}
\frac{d}{dx}[\tan(x)] &= \frac{d}{dx}\frac{\sin(x)}{\cos(x)} \\[2mm]
&= \frac{\cos(x)\frac{d}{dx}[\sin(x)] - \sin(x)\frac{d}{dx}[\cos(x)]}{[\cos(x)]^2} \\[2mm]
&= \frac{\cos(x)\cdot\cos(x) - \sin(x)\cdot[-\sin(x)]}{\cos^2(x)} \\[2mm]
&= \frac{\cos^2(x) + \sin^2(x)}{\cos^2(x)} \\[2mm]
&= \frac{1}{\cos^2(x)} \\[2mm]
&= \sec^2(x)
\end{aligned}
$$

All six basic trigonometric functions have derivatives, which are listed in Table 4.1. You should commit these derivatives to memory, as they will be used regularly throughout the course. Additionally, you will also need to recall values of the trigonometric functions at key radian measures such as multiples of $\dfrac{\pi}{6}$ and $\dfrac{\pi}{4}$. You should have learned these in a previous course; review them if you need to.

Table 4.1

Function	Derivative
$y = \sin(x)$	$\dfrac{dy}{dx} = \cos(x)$
$y = \cos(x)$	$\dfrac{dy}{dx} = -\sin(x)$
$y = \tan(x)$	$\dfrac{dy}{dx} = \sec^2(x)$
$y = \cot(x)$	$\dfrac{dy}{dx} = -\csc^2(x)$
$y = \sec(x)$	$\dfrac{dy}{dx} = \sec(x) \cdot \tan(x)$
$y = \csc(x)$	$\dfrac{dy}{dx} = -\csc(x) \cdot \cot(x)$

Close inspection shows that there are patterns in the derivatives that make them easier to learn. Notice that the derivative of each function beginning with "co" has a negative sign. Also notice that the derivatives of each *pair* of cofunctions are themselves cofunctions. For example, the derivative of tangent involves the secant function, and the derivative of *co*tangent involves the *co*secant function.

Watch for a couple of subtleties when working with and evaluating derivatives of trigonometric functions. The first regards the domain of each function. Only sine and cosine are defined for all real numbers; there are values of x for which the derivatives of the other four functions are not defined. The second thing to keep in mind is that you may be able to simplify trigonometric expressions that appear to involve very complicated differentiation for example, by applying some trigonometric identities prior to differentiation.

EXAMPLE 4.8

Find the equation of the line tangent to $f(x) = \tan(x)$ at $x = \dfrac{\pi}{4}$.

SOLUTION

$f\left(\dfrac{\pi}{4}\right) = \tan\left(\dfrac{\pi}{4}\right) = 1$, so the point at which the tangent is drawn is $\left(\dfrac{\pi}{4},\ 1\right)$.

$$f'(x) = \dfrac{d}{dx}[\tan(x)]$$
$$= \sec^2(x)$$

The slope of the tangent at $x = \dfrac{\pi}{4}$ is

$$\sec^2\left(\dfrac{\pi}{4}\right) = \dfrac{1}{\cos^2\left(\frac{\pi}{4}\right)}$$

$$= \dfrac{1}{\left(\frac{\sqrt{2}}{2}\right)^2}$$

$$= \dfrac{1}{\frac{1}{2}}$$

$$= 2$$

The equation in point-slope form is $y - 1 = 2\left(x - \dfrac{\pi}{4}\right)$.

EXAMPLE 4.9

If $y = \cot(x) \cdot \sec(x)$, find $\dfrac{dy}{dx}$.

SOLUTION

Rather than using the product rule, simplify the expression first.

$$\cot(x) \cdot \sec(x) = \dfrac{\cos(x)}{\sin(x)} \cdot \dfrac{1}{\cos(x)}$$

$$= \dfrac{1}{\sin(x)}$$

$$= \csc(x)$$

$$\frac{d}{dx}[\cot(x) \cdot \sec(x)] = \frac{d}{dx}[\csc(x)]$$

$$= -\csc(x) \cdot \cot(x)$$

EXAMPLE 4.10

On the interval $[0, 2\pi)$, where is $\cot(x)$ *not* differentiable?

SOLUTION

The function $\cot(x)$ is discontinuous and not differentiable where $\sin(x) = 0$. Also,

$$\frac{d}{dx}[\cot(x)] = -\csc^2(x)$$

$$= \frac{-1}{\sin^2(x)}$$

For the derivative as well, then, when $\sin(x) = 0$, the derivative is undefined. On the given interval $\sin(x) = 0$ at $x = 0$ and $x = \pi$.

4.5   DERIVATIVES OF EXPONENTIAL AND LOGARITHMIC FUNCTIONS

One of the most interesting functions in the field of math and science is the exponential function $y = e^x$. Exponential functions appear with great regularity in models related to population growth. This function is also fascinating because its instantaneous rate of change at any point is always equal to the value of the function at that point. In symbols, this means $\frac{dy}{dx} = y$, and substituting e^x for y gives $\frac{d}{dx} e^x = e^x$.

The graph of $f(x) = e^x$ is shown below with the line tangent to the function at the point $(0, 1)$. Notice that the slope of the line appears to be 1. This will be accepted without formal proof.

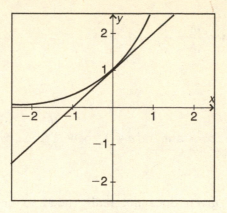

Figure 4.1

Recall the alternate definition of a derivative, $f'(a) = \lim\limits_{x \to a} \dfrac{f(x) - f(a)}{x - a}$. Let $a = 0$, replace x with h, and apply it to the function $f(x) = e^x$. Then the slope of the tangent to the graph at $x = 0$ is $\lim\limits_{h \to 0} \dfrac{e^h - 1}{h} = 1$. Now examine the derivative of e^x by definition.

$$
\begin{aligned}
\frac{d}{dx} e^x &= \lim_{h \to 0} \frac{e^{(x+h)} - e^x}{h} \\
&= \lim_{h \to 0} \frac{e^x e^h - e^x}{h} \\
&= \lim_{h \to 0} \frac{e^x (e^h - 1)}{h} \\
&= e^x \lim_{h \to 0} \frac{e^h - 1}{h} \\
&= e^x \cdot 1 \\
&= e^x
\end{aligned}
$$

$$\frac{d}{dx}(e^x) = e^x$$

Three more key derivatives are given here without proof.

If a is a positive constant and $a \neq 1$, then $\dfrac{d}{dx}(a^x) = a^x \cdot \ln(a)$.

If $y = \ln(x)$, then $\dfrac{dy}{dx} = \dfrac{1}{x}$.

If a is a positive constant and $a \neq 1$, then $\dfrac{d}{dx}(\log_a x) = \dfrac{1}{\ln(a)} \cdot \dfrac{1}{x}$.

The last derivative given does not really need to be memorized. It can be easily established by using the change of base rule for logarithms, $\log_a x = \dfrac{\ln(x)}{\ln(a)}$. Since $\ln(a)$ is a constant, taking the derivative of $\dfrac{\ln(x)}{\ln(a)}$ is no different than taking the derivative of $\ln(x)$ and multiplying the result by $\dfrac{1}{\ln(a)}$.

As with the trigonometric functions, the properties of exponential and logarithmic functions will be crucial to recall, especially in multiple-choice situations where a simplified choice of the derivative may be offered, as in Example 4.11.

EXAMPLE 4.11

If $f(x) = \log_3(x)$, then $f'(2) =$

(A) $\log_3 2$ (B) $\log_3 1$ (C) $\dfrac{1}{2\log(3)}$ (D) $\dfrac{1}{\ln(9)}$ (E) $\dfrac{1}{3\ln(2)}$

SOLUTION

$f(x) = \dfrac{1}{\ln(3)} \cdot \ln(x)$, so $f'(x) = \dfrac{1}{\ln(3)} \cdot \dfrac{1}{x}$, and $f'(2) = \dfrac{1}{2\ln(3)}$.

It seems that the correct solution is not offered, but (D) is the proper choice. Because of the property of logarithms,

$a \cdot \ln(b) = \ln(b^a)$, $\dfrac{1}{2\ln(3)} = \dfrac{1}{\ln(3^2)}$, which is $\dfrac{1}{\ln(9)}$.

This example shows the importance of knowing the properties of logarithms and exponential functions.

4.6 HIGHER-ORDER DERIVATIVES

If the derivative of a function exists and is itself differentiable, it is possible to take the derivative of the derivative. Naturally, this is called the second derivative. If the first derivative represents the instantaneous rate of change of the function at a point, the second derivative represents the rate of change of the rate of change. The significance of this idea will be explored in the next chapter. At this point, it is sufficient to master the notation and process of taking the second derivative, and possibly carrying that to higher orders of derivatives.

The second derivative is denoted in a variety of ways, the most common of which are y'', $f''(x)$, and $\dfrac{d^2 y}{dx^2}$. Similar notation carries to the third derivative and beyond with just a minor change. The prime marks on y and $f(x)$ eventually are replaced with numbers. So notation for the third derivative could be y''', $f'''(x)$, or may be seen as $y^{(3)}$, $f^{(3)}(x)$, or $\dfrac{d^3 y}{dx^3}$.

EXAMPLE 4.12

If $h(x) = \ln(x)$, find $h''(x)$.

SOLUTION

$$h'(x) = \frac{1}{x}, \text{ or } x^{-1}$$

$$h''(x) = -1x^{-2}, \text{ or } \frac{-1}{x^2}$$

4.7 CHAIN RULE

To this point, every function being differentiated was a simple function of x, such as $y = 3\sin(x)$, not a composite function such as $y = 3\sin(x^2)$. Composite functions require use of the chain rule. There is no rule in differential calculus more important and significant than the chain rule. It should be understood in multiple forms.

Chain Rule

> If function g is differentiable at x, h is differentiable at $g(x)$, and if $y = h(g(x))$ is a composite function of g and h, then
>
> $$\frac{dy}{dx} = h'(g(x)) \cdot g'(x).$$
>
> Using Leibniz notation, if $y = h(u)$ and $u = g(x)$, then
>
> $$\frac{dy}{dx} = \frac{dy}{du} \cdot \frac{du}{dx}.$$

Finding the correct derivative is often manageable when functions are presented in the Leibniz notation form. This is because the process identifies the functions that each need to be differentiated with respect to their independent variables.

EXAMPLE 4.13

If $y = \cos(x^2 + 1)$, find $\dfrac{dy}{dx}$.

SOLUTION

The function $y = \cos(x^2 + 1)$ is a composite of $y = \cos(u)$ and $u = x^2 + 1$.

$$\begin{aligned}
\frac{dy}{dx} &= \frac{dy}{du} \cdot \frac{du}{dx} \\
&= \frac{d}{du}[\cos(u)] \cdot \frac{d}{dx}(x^2 + 1) \\
&= -\sin(u) \cdot 2x \\
&= -2x \cdot \sin(x^2 + 1)
\end{aligned}$$

Notice the substitution for u to write the final answer as a function of x.

OUTSIDE-INSIDE PRINCIPLE

If the function is not already broken down, you have essentially two options. Break it down into a composite of two or more functions, as in the previous example, or learn to identify an outside and an inside function. If $y = h(g(x))$, is a composite function of g and h, think of h as the "outside" function and $g(x)$ as the "inside" function. Notice that if $\frac{dy}{dx} = f'(g(x)) \cdot g'(x)$, the derivative is found by differentiating the outside function, *without changing the inside function* $g(x)$, and then multiplying by the derivative of the inside function. Mastering this method can speed up the process of differentiation, but using it without having mastered the method will result in omitted chain rule factors.

Let us work Example 4.13 again by using the outside-inside principle.

EXAMPLE 4.14

If $y = \cos(x^2 + 1)$, find $\frac{dy}{dx}$.

SOLUTION

The outside function is the cosine function, and the inside function is the angle, $x^2 + 1$. The cosine will be differentiated without changing the angle, and then will be multiplied by the derivative of the inside function, $x^2 + 1$.

$$\frac{dy}{dx} = \frac{d}{dx}[\cos(x^2 + 1)]$$

$$= -\sin(x^2 + 1) \cdot \frac{d}{dx}(x^2 + 1)$$

$$= -\sin(x^2 + 1) \cdot 2x$$

$$= -2x \cdot \sin(x^2 + 1)$$

Remember, whenever a trigonometric function is differentiated, the angle will never be changed.

EXAMPLE 4.15

Find the value of x where the slope of the function $f(x) = e^{(x^2 - 2x)}$ equals 0.

SOLUTION

Think of the exponential function as the outside function and $x^2 - 2x$ as the inside function.

$$f'(x) = e^{(x^2 - 2x)} \cdot \frac{d}{dx}(x^2 - 2x)$$

$$= e^{(x^2 - 2x)} \cdot (2x - 2)$$

Now solve $e^{(x^2 - 2x)} \cdot (2x - 2) = 0$.

Either $e^{(x^2 - 2x)} = 0$ or $2x - 2 = 0$.

$e^{(x^2 - 2x)}$ is never 0, but $2x - 2 = 0$ when $x = 1$.

With the introduction of the chain rule as a way to find more complicated derivatives, it is important to keep in mind some of the rules learned earlier, such as the product rule and quotient rule. It is not uncommon to need to apply both the product rule and the chain rule in the same problem, as seen in the following example.

EXAMPLE 4.16

Find $\dfrac{dy}{dx}$ if $y = \ln(x^2 + 1) \cdot \tan(4x^3)$.

SOLUTION

The product rule is used first, and the derivatives in the product rule use the chain rule.

$$\frac{dy}{dx} = \ln(x^2 + 1) \cdot \frac{d}{dx}[\tan(4x^3)] + \tan(4x^3) \cdot \frac{d}{dx}[\ln(x^2 + 1)]$$

$$= \ln(x^2 + 1) \cdot \sec^2(4x^3) \cdot \frac{d}{dx}(4x^3) + \tan(4x^3) \cdot \frac{1}{x^2 + 1} \cdot \frac{d}{dx}(x^2 + 1)$$

$$= \ln(x^2 + 1) \cdot \sec^2(4x^3) \cdot 12x^2 + \tan(4x^3) \cdot \frac{1}{x^2 + 1} \cdot 2x$$

Sometimes the chain rule may need to be applied multiple times in a row within a problem. This occurs when the composition of more than two functions can be identified.

The following example shows both ways of handling a situation such as this. It is done by decomposition into individual functions and by multiple use of the outside-inside principle.

EXAMPLE 4.17

If $y = \sin^3(\sqrt{x})$, find $\dfrac{dy}{dx}$.

SOLUTION

Decompose the function into a cubic, a trigonometric, and a radical function.

Let $y = u^3$, $u = \sin(v)$, and $v = x^{\frac{1}{2}}$.

$$\frac{dy}{dx} = \frac{dy}{du} \cdot \frac{du}{dv} \cdot \frac{dv}{dx}$$

$$= 3u^2 \cdot \cos(v) \cdot \frac{1}{2} x^{-\frac{1}{2}}$$

$$= \frac{3}{2} \sin^2(v) \cos(v) \cdot \frac{1}{\sqrt{x}}$$

$$= \frac{3}{2\sqrt{x}} \sin^2(\sqrt{x}) \cos(\sqrt{x})$$

> Substitution back to the original variable is almost always required when using the chain rule.

It is sometimes easier to rewrite powers of trigonometric functions prior to using the outside-inside principle. Think of $\sin^3(\sqrt{x})$ as $[\sin(\sqrt{x})]^3$. The outermost function is the cubic function. The next function "in" is the sine function. And the innermost function is the radical.

$$\frac{d}{dx}[\sin(\sqrt{x})]^3 = 3[\sin(\sqrt{x})]^2 \cdot \frac{d}{dx}\sin(\sqrt{x})$$

$$= 3[\sin(\sqrt{x})]^2 \cdot \cos(\sqrt{x})\frac{d}{dx}(\sqrt{x})$$

$$= 3[\sin(\sqrt{x})]^2 \cdot \cos(\sqrt{x}) \cdot \frac{1}{2} x^{\left(-\frac{1}{2}\right)}$$

$$= \frac{3}{2\sqrt{x}} \sin^2(\sqrt{x}) \cos(\sqrt{x})$$

With the introduction of the chain rule into the course material, all the previously given derivative formulas take on a somewhat new look, each with the chain rule multiplier as a factor. Table 4.2 summarizes those formulas, each of which must be memorized. In all cases, u and v are differentiable functions of an unnamed independent variable; du is the chain rule derivative of that function; and k, n, and a are constants.

Table 4.2

Function	Derivative
$y = u^n$ (n is any real number)	$dy = n \cdot u^{(n-1)} \cdot du$
$y = kn$ (k is a constant)	$dy = 0$
$y = u \cdot v$	$dy = u \cdot dv + v \cdot du$
$y = \dfrac{u}{v}$	$dy = \dfrac{v \cdot du - u \cdot dv}{v^2}$
$y = f(g(x))$	$\dfrac{dy}{dx} = f'(g(x)) \cdot g'(x)$
$y = \sin(u)$	$dy = \cos(u) \cdot du$
$y = \cos(u)$	$dy = -\sin(u) \cdot du$
$y = \tan(u)$	$dy = \sec^2(u) \cdot du$
$y = \cot(u)$	$dy = -\csc^2(u) \cdot du$
$y = \sec(u)$	$dy = \sec(u) \cdot \tan(u)\, du$
$y = \csc(u)$	$dy = -\csc(u) \cdot \cot(u)\, du$
$y = e^u$	$dy = e^u \cdot du$
$y = a^u$	$dy = a^u \cdot \ln(a) \cdot du$
$y = \ln(u)$	$dy = \dfrac{du}{u}$
$y = \log_a u$	$dy = \dfrac{1}{\ln(a)} \cdot \dfrac{du}{u}$

HIGHER-ORDER DERIVATIVES WITH THE CHAIN RULE

Higher-order derivatives should now be briefly reexamined in light of the chain rule. Because the chain rule introduces new factors into the first derivative, finding a second derivative may now require using the product or quotient rule. It is important to be alert to this possibility.

EXAMPLE 4.18

Find $\dfrac{d^2 g}{dx^2}$ if $g(x) = \ln(x^2 + 1)$.

SOLUTION

$$\frac{dg}{dx} = \frac{1}{x^2 + 1} \cdot \frac{d}{dx}(x^2 + 1)$$

$$= \frac{2x}{x^2 + 1}$$

The next derivative now requires the use of the quotient rule.

$$\frac{d^2 g}{dx^2} = \frac{d}{dx}\left(\frac{2x}{x^2 + 1}\right)$$

$$= \frac{(x^2 + 1)\frac{d}{dx}(2x) - (2x)\frac{d}{dx}(x^2 + 1)}{(x^2 + 1)^2}$$

$$= \frac{(x^2 + 1)(2) - 2x(2x)}{(x^2 + 1)^2}$$

$$= \frac{-2x^2 + 2}{(x^2 + 1)^2}$$

EXAMPLE 4.19

If $f(x) = e^{3x}$, find an expression for $f^{(n)}(x)$, the nth derivative of f.

SOLUTION

$$f'(x) = e^{3x} \cdot \frac{d}{dx}(3x)$$

$$= 3e^{3x}$$

$$f''(x) = 3\left[e^{3x} \cdot \frac{d}{dx}(3x) \right]$$

$$= 9e^{3x}$$

By observation of the pattern, the repeated chain rule factor of the derivative of $3x$ provides one more factor of 3 each time differentiation occurs:

$$f^{(n)}(x) = 3^n \cdot e^{3x}.$$

4.8 IMPLICIT DIFFERENTIATION

There are times when it is either difficult or impossible to write a function in the form $y = f(x)$, with y completely isolated from x. Functions and relations such as this are called implicit. Fortunately, the chain rule still enables mathematicians to find the derivative, which itself may be an implicit combination of both variables x and y! There are two very subtle aspects to differentiating implicit functions. The first thing to remember is that even though y cannot be isolated, it is still considered a function of x. The practical result of this is that whenever a y term is differentiated, it will produce a $\frac{dy}{dx}$ chain rule factor. The second subtlety is remembering to use the product or quotient rule whenever a product or quotient of x and y is encountered. For instance, in the implicit equation $x^2 y^2 + 2x = y^3$, the term $x^2 y^2$ is a product of two functions of x, x^2 and y^2, and would require the use of the product rule. One way to gain confidence in the process is to examine an example that can be differentiated in either its explicit or its implicit form and see that the results are indeed the same.

EXAMPLE 4.20

Given $\sin(x) \cdot y = 4$, find $\frac{dy}{dx}$.

SOLUTION

Isolating y leads to $y = \dfrac{4}{\sin(x)}$ or $y = 4\csc(x)$, so $\dfrac{dy}{dx} = -4\csc(x) \cdot \cot(x)$.

Now we use implicit differentiation with the product rule.

$$\frac{d}{dx}[\sin(x) \cdot y] = \frac{d}{dx}(4)$$

$$\sin(x) \cdot \frac{d}{dx}(y) + y \cdot \frac{d}{dx}[\sin(x)] = 0$$

$$\sin(x)\frac{d}{dx} + y \cdot \cos(x) = 0$$

Solving for $\frac{dy}{dx}$ yields $\frac{dy}{dx} = \frac{-y \cdot \cos(x)}{\sin x}$.

$$\frac{dy}{dx} = -y \cdot \cot(x)$$

Since $y = 4\csc(x)$, substituting for y produces $\frac{dy}{dx} = -4\csc(x)\cot(x)$.

Now let's examine the equation given in the previous paragraph to see how the product rule works in implicit differentiation.

EXAMPLE 4.21

Given $x^2y^2 + 2x = y^3$, find $\frac{dy}{dx}$ in terms of x and y.

SOLUTION

$$\frac{d}{dx}(x^2y^2 + 2x) = \frac{d}{dx}(y^3)$$

$$\frac{d}{dx}(x^2y^2) + \frac{d}{dx}(2x) = \frac{d}{dx}(y^3)$$

> Notice the use of the product rule on the left side of the equation *and* the power rule on the right side of the equation.

$$x^2 \cdot \frac{d}{dx}(y^2) + y^2 \frac{d}{dx}(x^2) + 2 = 3y^2 \cdot \frac{d}{dx}(y)$$

$$x^2 \cdot 2y\frac{d}{dx}(y) + y^2 \cdot 2x + 2 = 3y^2 \cdot \frac{dy}{dx}$$

The $\dfrac{dy}{dx}$ term is produced by differentiating the base of each power of y. It is the chain rule factor.

$$2x^2 y\, \frac{dy}{dx} + 2xy^2 + 2 = 3y^2\, \frac{dy}{dx}$$

$$2xy^2 + 2 = 3y^2\, \frac{dy}{dx} - 2x^2 y\, \frac{dy}{dx}$$

$$2xy^2 + 2 = (3y^2 - 2x^2 y)\, \frac{dy}{dx}$$

$$\frac{2xy^2 + 2}{3y^2 - 2x^2 y} = \frac{dy}{dx}$$

Notice that the derivative value now relies on both the x and y values at a given point. Sometimes when finding the value of the derivative at a point, one may have to actually use the entire ordered pair instead of just the abscissa.

EXAMPLE 4.22

Find the equation of the line tangent to $x^2 + \dfrac{x}{y} = 2y$ in the first quadrant at $y = 1$.

SOLUTION

First solve for x when $y = 1$.

$$x^2 + \frac{x}{1} = 2\cdot 1$$

$$x^2 + x - 2 = 0$$

$$(x + 2)(x - 1) = 0$$

$$x = -2 \text{ or } x = 1$$

Keep $x = 1$ because the problem requires a first quadrant point. The point is $(1, 1)$.

Find $\dfrac{dy}{dx}$ to get the slope. The term $\dfrac{x}{y}$ requires using the quotient rule.

$$\frac{d}{dx}(x^2) + \frac{d}{x}\left(\frac{x}{y}\right) = \frac{d}{dx}(2y)$$

$$2x + \left(\frac{y \cdot 1 - x\frac{dy}{dx}}{y^2}\right) = 2\frac{dy}{dx}$$

Multiply each term by y^2.

$$2xy^2 + y - x\frac{dy}{dx} = 2y^2\frac{dy}{dx}$$

$$\frac{dy}{dx} = \frac{2xy^2 + y}{2y^2 + x}\bigg|_{(1,1)} = \frac{3}{3} = 1$$

The point-slope form of the equation is

$$y - 1 = 1(x - 1), \text{ or } y = x$$

The implicit curve and the line tangent at (1, 1) are shown in Figure 4.2.

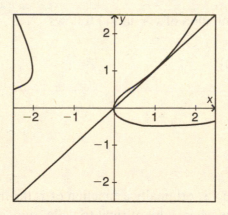

Figure 4.2

4.9   DERIVATIVES OF INVERSE TRIGONOMETRIC FUNCTIONS

Implicit differentiation enables the exploration of the derivatives of an additional set of functions, inverse trigonometric functions. As with the normal trigonometric functions, there are six inverses. The first derivative will be justified, and the rest will be presented without proof. Once again, these derivatives need to be committed to memory, but fortunately, they have cofunction patterns that will ease the task.

Remember, the domain values of the inverse trigonometric functions are trigonometric ratios, real numbers, and the range consists of radian measures with appropriately defined limitations. The function $y = \arctan(x)$ essentially says, "Tell me the angle y with a tangent value of x." (Note: Unless otherwise noted, $\arctan(x)$, will be also written with the notation $\tan^{-1}(x)$. This notation is also used with the other inverse functions, and should not be thought of as a power of -1.)

$\dfrac{dy}{dx}$ for $y = \tan^{-1}(x)$ denotes the instantaneous rate of change of the radian measure as the trigonometric ratio changes value. It also finds an expression for the slope of the graph of the inverse function at any given point of its domain.

To find $\dfrac{dy}{dx}$ for $y = \tan^{-1}(x)$, we write an equivalent expression, $x = \tan(y)$.

The implicit differentiation of $x = \tan(y)$ follows.

$$\frac{d}{dx}(x) = \frac{d}{dx}[\tan(y)]$$
$$1 = \sec^2(y) \cdot \frac{dy}{dx}$$
$$\frac{dy}{dx} = \cos^2(y)$$

Unfortunately, since y is originally a function of x, the derivative also should be in terms of x. This is where a bit of right triangle trigonometry is helpful!

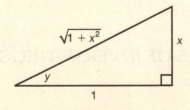

Figure 4.3

Figure 4.3 shows a right triangle with an acute angle y and the legs labeled appropriately so that $\tan(y) = x$. By finding the hypotenuse, any of the other trigonometric ratios for y are available.

$$\cos(y) = \frac{1}{\sqrt{1+x^2}} \text{, so } \cos^2(y) = \frac{1}{1+x^2}.$$

If $y = \tan^{-1}(x)$, then $\dfrac{dy}{dx} = \dfrac{1}{1+x^2}$.

This result can be furthered generalized by using the chain rule.

If u is a function of x, and $y = \tan^{-1}(u)$, then $\dfrac{dy}{du} = \dfrac{1}{1+u^2} \cdot \dfrac{du}{dx}$. Even more generally, if u is a function of any unnamed variable, then $d(\tan^{-1}(u)) = \dfrac{du}{1+u^2}$.

Table 4.3 summarizes the derivatives of the six inverse trigonometric functions. In all cases, u is a differentiable function of an unnamed independent variable, and du is the chain rule derivative of that function. Notice that the form for each cofunction differs only by a negative sign from that for the corresponding function, so there are actually only three that must be memorized.

Table 4.3

Function	Derivative
$y = \tan^{-1}(u)$	$dy = \dfrac{du}{1+u^2}$
$y = \cot^{-1}(u)$	$dy = \dfrac{-du}{1+u^2}$
$y = \sin^{-1}(u)$	$dy = \dfrac{du}{\sqrt{1-u^2}}$
$y = \cos^{-1}(u)$	$dy = \dfrac{-du}{\sqrt{1-u^2}}$
$y = \sec^{-1}(u)$	$dy = \dfrac{du}{|u|\sqrt{u^2-1}}$
$y = \csc^{-1}(u)$	$dy = \dfrac{-du}{|u|\sqrt{u^2-1}}$

EXAMPLE 4.23

Find $\dfrac{dy}{dz}$ if $y = \sec^{-1}(3z)$ for $z > \dfrac{1}{3}$.

SOLUTION

Think of $3z$ as u in the formula for the derivative, $d(\sec^{-1} u) = \dfrac{du}{|u|\sqrt{u^2 - 1}}$.

$$\frac{dy}{dz} = \frac{\frac{d}{dz}(3z)}{|3z|\sqrt{(3z)^2 - 1}}$$

$$\frac{dy}{dz} = \frac{3}{|3z|\sqrt{9z^2 - 1}}$$

EXAMPLE 4.24

Find any x values where the graph of $f(x) = \cot^{-1}(x^2)$ has a horizontal tangent.

SOLUTION

A horizontal tangent has a slope of zero, so the task is to find where $f'(x) = 0$.

$$\frac{d}{dx}[\cot^{-1}(x^2)] = \frac{-\frac{d}{dx}(x^2)}{1 + (x^2)^2}$$

$$= \frac{-2x}{1 + x^4}$$

$\dfrac{-2x}{1 + x^4} = 0$ when $x = 0$.

Since the domain of the function is all real numbers, and $f'(0)$ is defined, the graph has a horizontal tangent at $x = 0$.

4.10 DERIVATIVES OF INVERSE FUNCTIONS

Recall that the graphs of inverse functions, whether trigonometric or not, are reflections across the line $y = x$. This geometric relationship has interesting ramifications for the slopes of the tangents to the curves at image points (the point on one graph that is the reflection of a point on its inverse graph). Fig. 4.4

shows the graphs of $f(x) = x^2 + 1$ for $x \geq 0$ and its inverse $g(x) = \sqrt{x-1}$. The lines tangent to f at $(1, 2)$ and g at the image point $(2, 1)$ are also shown.

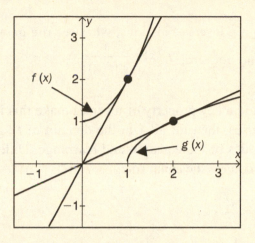

Figure 4.4

Observe that the slopes of the two tangent lines are reciprocals, 2 and $\dfrac{1}{2}$.

This, of course, is no coincidence. The same thing will happen at image points on any two functions that are inverses. If the tangent line to one function is already drawn prior to reflecting across the line $y = x$, when the reflection is done to generate the inverse function, the original tangent line is also reflected. The new tangent to the inverse function is actually the inverse of the original tangent. So how do the slopes of two lines that are reflections across $y = x$ compare?

Let L_1 be any line with the equation $y = mx + b$. To find the inverse function, switch the x and y, and isolate y again.

$x = my + b$

$my + b = x$

$my = x - b$

$y = \dfrac{x-b}{m}$

$y = \left(\dfrac{1}{m}\right)x - \dfrac{b}{m}$

The slope of L_1 is m, and the slope of its inverse is the reciprocal, $\dfrac{1}{m}$.

> If f and g are inverse functions, wherever the derivatives are defined and not zero, $f'(x) = \dfrac{1}{g'[f(x)]}$.

The chain rule and a key property of inverses make this fact easy to support. If f and g are inverses, then for all x in the domain of f, $g(f(x)) = x$. Simply differentiate both sides of the equation and rearrange. Differentiating the composite of inverses requires the chain rule.

$$\frac{d}{dx}[g(f(x))] = \frac{d}{dx}(x)$$
$$g'(f(x)) \cdot f'(x) = 1$$
$$f'(x) = \frac{1}{g'(f(x))}$$

Notice that the derivative of g is evaluated at $f(x)$, which is the y-coordinate on the graph of f. The most common mistake students make when using this property is attempting to evaluate each derivative at the same value. When inverses are created analytically, the *first step* is to switch the x and y. This causes the range of the original function to become the domain of its inverse. Pay special attention to this detail. Fig. 4.5 is provided to give visual reinforcement to this subtle aspect of derivatives of inverse functions. f and g in the figure are generic inverse functions.

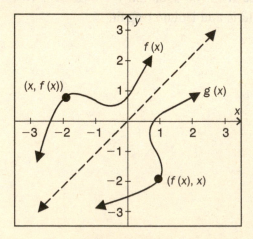

Figure 4.5

EXAMPLE 4.25

If g and h are inverses, with $h(3) = 7$ and $g'(7) = \dfrac{1}{4}$, find the equation of the line tangent to h at $x = 3$.

SOLUTION

The slope of the tangent is $h'(3)$. $h'(3) = \dfrac{1}{g'(h(3))}$, so

$$h'(3) = \frac{1}{g'(7)} = 4 \,.$$

The equation of the tangent line is thus $y - 7 = 4(x - 3)$, or $y = 4x - 5$.

4.11 EXERCISES

Find $\dfrac{dy}{dx}$ for the first 15 problems below. This should take less than 10 minutes. Then continue solving the remaining problems.

1. $y = \sin(x)$

2. $y = 3^x$

3. $y = \ln(x + 2)$

4. $y = e^{\tan(x)}$

5. $y = \sec(x)$

6. $y = \dfrac{4}{x^2}$

7. $y = \cos^{-1}(4x)$

8. $y = \log_3(x^2 + 1)$

9. $y = x^3 \cdot \cot(x)$

10. $y = f(h(x))$

11. $y = (2x^3 - 4x)^5$

12. $y = \dfrac{4x^2 - 9}{2x + 3}$

13. $y = \sec^{-1}(3x)$

14. $y = \csc(x)\tan(x)$

15. $y = \dfrac{4}{5^x}$

16. Find $\dfrac{dy}{dx}$ if $y = u^2$ and $u = \csc(8x)$

17. Find the equation of the line tangent to the graph of $x^2y^2 - y^2 = xy - 4$ at $x = 1$.

Table 4.4

x	$u(x)$	$v(x)$	$u'(x)$	$v'(x)$
2	−3	4	6	5
3	2	7	−2	1

18. If $f(x) = \dfrac{u(x)}{v(x)}$, use Table 4.4 to find $f'(2)$.

19. If $y = v(u(x))$, use Table 4.4 to find $\dfrac{dy}{dx}$ when $x = 3$.

20. Find $\dfrac{d^2y}{dx^2}$ for $y = \sin(x^3 + 1)$.

21. The distance of a bike from a starting position is given by $s(t) = \dfrac{2t + 1}{4\sqrt{t + 1}}$. If s is measured in feet and t is in seconds, find the velocity of the bike at $t = 3$ seconds.

22. If f and g are inverse functions with nonzero derivatives at all real numbers, what is the value of $g'(f(x)) \cdot f'(x)$?

23. Find all values of x where $h(x) = \tan^{-1}(x^2 + 2x)$ has horizontal tangents.

24. Given $y = h(u)$ and $u = g(x)$ are continuous and differentiable functions, if $\dfrac{dy}{du} = \dfrac{2}{3}$ and $\dfrac{dy}{dx} = 12$, find $\dfrac{du}{dx}$.

25. Let $h(x)$ be the inverse of $f(x)$, and $f(x) = x^3 + 2x + 1$. Find $h'(1)$.

26. Find the nth derivative of $p(x) = x^n$.

4.12 SOLUTIONS TO EXERCISES

1. $\dfrac{dy}{dx} = \cos(x)$

2. $\dfrac{dy}{dx} = 3^x \cdot \ln(3)$

3. $\dfrac{dy}{dx} = \dfrac{1}{x+2}$

$\dfrac{dy}{dx} = \dfrac{\frac{dy}{dx}(x+2)}{x+2}$

$= \dfrac{1}{x+2}$

4. $\dfrac{dy}{dx} = \sec^2(x) \cdot e^{\tan(x)}$

$\dfrac{dy}{dx} = \dfrac{d}{dx}[e^{\tan(x)}]$

$= e^{\tan(x)} \cdot \dfrac{d}{dx}[\tan(x)]$

$= \sec^2(x) \cdot e^{\tan(x)}$

5. $\dfrac{dy}{dx} = \sec(x) \cdot \tan(x)$

6. $\dfrac{dy}{dx} = \dfrac{-8}{x^3}$

   $y = \dfrac{1}{x^2} = 4x^{-2}$

   $\dfrac{dy}{dx} = 4 \cdot -2x^{-2-1}$

   $\quad = -8x^{-3}$

   $\quad = \dfrac{-8}{x^3}$

7. $\dfrac{dy}{dx} = \dfrac{-4}{\sqrt{1-16x^2}}$

   $\dfrac{dy}{dx} = \dfrac{d}{dx}\cos^{-1}(4x)$

   $\quad = \dfrac{-\frac{d}{dx}(4x)}{\sqrt{1-(4x)^2}}$

   $\quad = \dfrac{-4}{\sqrt{1-16x^2}}$

8. $\dfrac{dy}{dx} = \dfrac{2x}{(x^2+1)\ln(3)}$

   $\dfrac{dy}{dx} = \dfrac{d}{dx}[\log_3(x^2+1)]$

   $\quad = \dfrac{1}{\ln(3)} \cdot \dfrac{\frac{d}{dx}(x^2+1)}{x^2+1}$

   $\quad = \dfrac{1}{\ln(3)} \cdot \dfrac{2x}{x^2+1}$

9. $\dfrac{dy}{dx} = -x^3 \csc^2(x) + 3x^2 \cot(x)$

$$\frac{dy}{dx} = \frac{d}{dx}[x^3 \cdot \cot(x)]$$

$$\frac{dy}{dx} = x^3 \cdot \frac{d}{dx}[\cot(x)] + \cot(x) \cdot \frac{d}{dx}(x^3)$$

$$= x^3 \cdot [-\csc^2(x)] + \cot(x) \cdot 3x^2$$

$$= -x^3 \csc^2(x) + 3x^2 \cot(x)$$

10. $\dfrac{dy}{dx} = f'(h(x)) \cdot h'(x)$

11. $\dfrac{dy}{dx} = (30x^2 - 20)(2x^3 - 4x)^4$

$$\frac{dy}{dx} = \frac{d}{dx}(2x^3 - 4x)^5$$

$$= 5(2x^3 - 4x)^{5-1} \cdot \frac{d}{dx}(2x^3 - 4x)$$

$$= 5(2x^3 - 4x)^4(6x^2 - 4)$$

$$= (30x^2 - 20)(2x^3 - 4x)^4$$

12. $\dfrac{dy}{dx} = 2$ Simplify before differentiating; otherwise, the quotient rule will require a lot of unnecessary work.

$$y = \frac{4x^2 - 9}{2x + 3}$$

$$= \frac{(2x - 3)(2x + 3)}{2x + 3}$$

$$y = 2x - 3$$

$$\frac{dy}{dx} = 2$$

13. $\dfrac{dy}{dx} = \dfrac{3}{|3x|\sqrt{9x^2-1}}$

$$\dfrac{dy}{dx} = \dfrac{d}{dx}[\sec^{-1}(3x)]$$

$$= \dfrac{\frac{d}{dx}(3x)}{|3x|\sqrt{(3x)^2-1}}$$

$$= \dfrac{3}{|3x|\sqrt{9x^2-1}}$$

14. $\dfrac{dy}{dx} = \sec(x)\tan(x)$

Simplify the trigonometric expression before differentiating.

$$y = \csc(x)\tan(x)$$

$$= \dfrac{1}{\sin(x)} \cdot \dfrac{\sin(x)}{\cos(x)}$$

$$= \dfrac{1}{\cos(x)}$$

$$= \sec(x)$$

$$\dfrac{dy}{dx} = \dfrac{d}{dx}[\sec(x)]$$

$$= \sec(x)\tan(x)$$

15. $\dfrac{dy}{dx} = \dfrac{-4\ln(5)}{5^x}$

The numerator is a constant, so the quotient rule is unnecessary.

$$y = \dfrac{4}{5^x} = 4 \cdot 5^{(-x)}$$

$$\dfrac{dy}{dx} = 4\dfrac{d}{dx}\left(5^{(-x)}\right)$$

$$= 4 \cdot 5^{(-x)} \cdot \ln(5) \cdot \dfrac{d}{dx}(-x)$$

$$= \dfrac{4\ln(5)}{5^x} \cdot (-1)$$

$$= \dfrac{-4\ln(5)}{5^x}$$

16. $\dfrac{dy}{dx} = -16 \csc^2(8x)\cot(8x)$

$\dfrac{dy}{dx} = 2u$

$\dfrac{du}{dx} = \dfrac{d}{dx}[\csc(8x)]$

$\quad = -\csc(8x)\cot(8x)\dfrac{d}{dx}(8x)$

$\quad = -8\csc(8x)\cot(8x)$

$\dfrac{dy}{dx} = \dfrac{dy}{du}\cdot\dfrac{du}{dx}$

$\quad = 2u[-8\csc(8x)\cot(8x)]$

$\quad = 2\csc(8x)\cdot[(-8)\csc(8x)\cot(8x)]$

$\quad = -16\csc^2(8x)\cot(8x)$

17. $y - 1 = 28(x - 1)$, or $y = 28x - 24$

Find the corresponding y-value when $x = 1$

$1^2 y^2 - y^2 = 1y - 4$

$\qquad 0 = y - 4$

$\qquad y = 4$

Use implicit differentiation to find $\dfrac{dy}{dx}$. Remember the product rule.

$\dfrac{d}{dx}(x^2 y^2 - y^2) = \dfrac{d}{dx}(xy - 4)$

$x^2\dfrac{d}{dx}(y^2) + y^2\dfrac{d}{dx}(x^2) - \dfrac{d}{dx}(y^2) = x\dfrac{d}{dx}(y) + y\dfrac{d}{dx}(x) - \dfrac{d}{dx}(4)$

$x^2\cdot 2y\dfrac{dy}{dx} + y^2\cdot 2x - 2y\dfrac{dy}{dx} = x\dfrac{dy}{dx} + y\cdot 1 - 0$

Substitute (1, 4) to find the value of $\dfrac{dy}{dx}$.

$$1^2 \cdot 2 \cdot 4 \frac{dy}{dx} + 4^2 \cdot 2 \cdot 1 - 2 \cdot 4 \frac{dy}{dx} = 1 \cdot \frac{dy}{dx} + 4 \cdot 1$$

$$8 \frac{dy}{dx} + 32 - 8 \frac{dy}{dx} = \frac{dy}{dx} + 4$$

$$\frac{dy}{dx} = 28$$

Write the equation using the point and the slope.

$$y - 4 = 28(x - 1)$$

18. $f'(2) = \dfrac{39}{16}$

Use the quotient rule, $f'(x) = \dfrac{v(x) \cdot u'(x) - u(x) \cdot v'(x)}{[v(x)]^2}$ and substitute from Table 4.4.

$$f'(2) = \frac{v(2) \cdot u'(2) - u(2) \cdot v'(2)}{[v(2)]^2}$$

$$= \frac{4(6) - (-3)(5)}{4^2}$$

$$= \frac{39}{16}$$

19. $\dfrac{dy}{dx} = -10$

Use the chain rule, $\dfrac{dy}{dx} = v'(u(x)) \cdot u'(x)$ and substitute from Table 4.4.

$$\frac{dy}{dx}\bigg|_{x-3} = v'(u(3)) \cdot u'(3)$$

$$= v'(2) \cdot u'(3)$$

$$= 5 \cdot (-2)$$

$$= -10$$

20. $\dfrac{d^2 y}{dx^2} = -9x^4 \sin(x^3 + 1) + 6x \cos(x^3 + 1)$

$\dfrac{dy}{dx} = \dfrac{d}{dx}[\sin(x^3 + 1)]$

$= \cos(x^3 + 1) \cdot \dfrac{d}{dx}(x^3 + 1)$

$= \cos(x^3 + 1) \cdot 3x^2$

Now use the product rule and the chain rule on the $\cos(x^3 + 1)$ term.

$\dfrac{d}{dx}\left(\dfrac{dy}{dx}\right) = \dfrac{d}{dx}[3x^2 \cdot \cos(x^3 + 1)]$

$\dfrac{d^2 y}{dx^2} = 3x^2 \cdot \dfrac{d}{dx}[\cos(x^3 + 1)] + [\cos(x^3 + 1)] \cdot \dfrac{d}{dx}(3x^2)$

$= 3x^2 \cdot [-\sin(x^3 + 1)] \cdot \dfrac{d}{dx}(x^3 + 1) + \cos(x^3 + 1) \cdot 6x$

$= 3x^2 \cdot [-\sin(x^3 + 1)] \cdot 3x^2 + 6x \cdot \cos(x^3 + 1)$

$= -9x^4 \sin(x^3 + 1) + 6x \cdot \cos(x^3 + 1)$

21. $v(3) = \dfrac{9}{64}$ ft/sec

The derivative of position is velocity. Differentiate and evaluate at $t = 3$

$\dfrac{ds}{dt} = \dfrac{d}{dt}\left(\dfrac{2t+1}{4\sqrt{t+1}}\right)$

$= \dfrac{1}{4} \cdot \dfrac{d}{dt}\left(\dfrac{2t+1}{\sqrt{t+1}}\right)$

$= \dfrac{1}{4} \cdot \dfrac{\sqrt{t+1}\,\frac{d}{dt}(2t+1) - (2t+1)\frac{d}{dt}\sqrt{t+1}}{(\sqrt{t+1})^2}$

$= \dfrac{1}{4} \cdot \dfrac{\sqrt{t+1}\cdot 2 - (2t+1)\frac{1}{2}(t+1)^{-\frac{1}{2}}}{t+1}$

$= \dfrac{2\sqrt{t+1} - \frac{2t+1}{2\sqrt{t+1}}}{4(t+1)}$

At $t = 3$.

$$\frac{ds}{dt} = \frac{2\sqrt{3+1} - \frac{2 \cdot 3 + 1}{2\sqrt{3+1}}}{4(3+1)}$$

$$= \frac{4 - \frac{7}{4}}{16}$$

$$= \frac{9}{64} \text{ ft/sec}$$

22. 1

If f and g are inverses, $g(f(x)) = x$, so $\frac{d}{dx}[g(f(x))] = \frac{d}{dx}(x)$, and $g'(f(x)) \cdot f'(x) = 1$.

23. $x = -1$

A horizontal tangent has a slope of 0. Solve $h'(x) = 0$.

If $h(x) = \tan^{-1}(x^2 + 2x)$, then $h'(x) = \frac{d}{dx}[\tan^{-1}(x^2 + 2x)]$.

$$\frac{d}{dx}[\tan^{-1}(x^2 + 2x)] = \frac{\frac{d}{dx}(x^2 + 2x)}{1 + (x^2 + 2x)^2}$$

$$= \frac{2x + 2}{1 + (x^2 + 2x)^2}$$

$\dfrac{2x + 2}{1 + (x^2 + 2x)^2} = 0$ when $2x + 2 = 0$ and the denominator is not 0.

$1 + (x^2 + 2x)^2 \geq 1$ for all x, so $2x + 2 = 0$ means $x = -1$.

24. $\dfrac{du}{dx} = 18$

By the chain rule, $\dfrac{dy}{dx} = \dfrac{dy}{du} \cdot \dfrac{du}{dx}$.

Substituting gives $12 = \dfrac{2}{3} \cdot \dfrac{du}{dx}$

$$\frac{du}{dx} = 12 \cdot \frac{3}{2} = 18$$

25. $h'(1) = \dfrac{1}{2}$

Fig. 4.5 shows $f(x)$, $h(x)$, and the points where the slopes are being calculated.

$$f'(x) = \frac{1}{h'(f(x))} \text{ so } h'(f(x)) = \frac{1}{f'(x)}.$$

$f(x) = 1$ when $x = 0$.

$$h'(1) = \frac{1}{f'(0)}$$

$$f'(x) = 3x^2 + 2$$

$$f'(0) = 2, \text{ so } h'(1) = \frac{1}{2}.$$

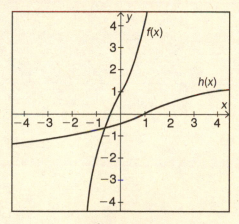

Figure 4.5

26. $p^{(n)}(x) = n!$

Examine the pattern of derivatives as $p(x) = x^n$ is repeatedly differentiated.

$$p'(x) = n \cdot x^{n-1}$$

$$p''(x) = n \cdot (n-1)x^{n-2}$$

$$p^{(3)}(x) = n(n-1)(n-2)x^{n-3}$$

$$p^{(4)}(x) = n\,(n-1)(n-2)\,(n-3)x^{n-4}$$

$$p^{(n)}(x) = n(n-1)(n-2)\ldots[n-(n-1)] \cdot x^{n-n}$$

$$p^{(n)}(x) = n(n-1)(n-2)\ldots(1) \cdot x^0 = n!$$

CHAPTER 5

Applications of Differentiation

CHAPTER 5

APPLICATIONS OF DIFFERENTIATION

5.1 INTRODUCTION

As with all mathematical concepts, the power of differentiation comes from applying its concepts either to expanding our understanding of further mathematics or to explaining observed real-life situations. Derivatives can be used for both of these pursuits. Since they allow us to measure change—both instantaneous and average rates of change—and since functions and the world around us are not static, this chapter applies derivatives to the study of function behavior and real-world applications.

5.2 FUNCTIONS AND FIRST-DERIVATIVE APPLICATIONS

LOCAL EXTREME VALUES

One of the most important applications of calculus is determining maximum or minimum values for situations that involve many options. For example, manufacturers desire to maximize profits on items they produce. Profits will fluctuate depending on a variety of variables, such as overhead costs, selling price, or market conditions. In industry, the problem often is analyzed with numerical models, or by using multivariable calculus. The foundations of these processes begin with simple optimization, as experienced in this course. Set as goals the development of both an intuitive understanding of maxima and minima and a mastery of the analytic work behind determining them.

Figure 5.1 shows examples of local maxima and minima and the various ways they can occur on the graph of a function.

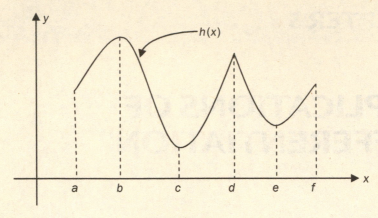

Figure 5.1

The abscissa is considered the location of a local maximum or minimum, and the ordinate is the actual maximum or minimum value. For example, a correct statement is, "The graph of $h(x)$ has a local maximum of $h(b)$ at $x = b$." Informally, a local maximum is the highest point in a small interval, and a local minimum is the lowest point in a small interval. In Figure 5.1, local maxima exist at b, d, and f. Local minima occur at a, c, and e. Notice, with the exception of the endpoints, that local extrema (high or low points) exist where the slope of the tangent is zero (b, c, and e), or undefined (d). These are called critical points.

Critical Point

> A critical point on a function h is a point $x = c$ in its domain where $h'(c) = 0$ or $h'(c)$ is undefined.

On the interior of the domain of a function, critical points are the candidates for local extrema but are not guaranteed to be maxima or minima. Simple graphical examples will be provided here as evidence. Figure 5.2 shows a parabola with its vertex at $(2, -1)$. At that point, a critical point, the derivative has a value of 0, and a local minimum exists. Figure 5.3 shows a cubic function with a plateau at $(1, 1)$. The value of the derivative at $(1, 1)$ is 0, making it a critical point. But even though the derivative at that point has a value of 0, no local maximum or minimum exists there.

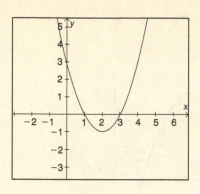

Figure 5.2

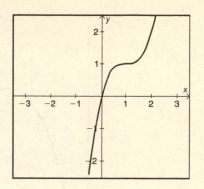

Figure 5.3

EXAMPLE 5.1

Find the critical values for $f(x) = x^{\left(\frac{2}{3}\right)} - x^{\left(\frac{8}{3}\right)}$.

SOLUTION

$$f'(x) = \frac{2}{3} x^{\left(\frac{-1}{3}\right)} - \frac{8}{3} x^{\left(\frac{5}{3}\right)}$$

$$\frac{2}{3} x^{\left(\frac{-1}{3}\right)} - \frac{8}{3} x^{\left(\frac{5}{3}\right)} = 0$$

$$\frac{2}{3} x^{\left(\frac{-1}{3}\right)} (1 - 4x^2) = 0$$

$$1 - 4x^2 = 0, \text{ or } \frac{1}{x^{\left(\frac{1}{3}\right)}} = 0$$

These two equations yield $f'(x) = 0$ at $x = \pm\frac{1}{2}$, and $f'(x)$ is undefined at $x = 0$.

The critical points are therefore $\pm\frac{1}{2}$ and 0.

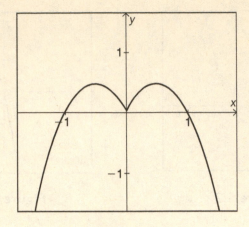

Figure 5.4

Figure 5.4 is the graph for $f(x)$. It shows that extreme values do exist at the critical points. There are local maxima at $x = \pm \dfrac{1}{2}$, and a local minimum at $x = 0$.

ABSOLUTE EXTREME VALUES

In addition to being able to determine the highest and lowest points on limited intervals of the domain, it is also important to be able to determine the absolute highest or lowest values of a function if those values exist. Critical points are the first step in the process, since local maxima and minima become candidates for absolute extremes. The only other candidates are the endpoints of the domain. To determine an absolute extreme for a function, compare the function values at the critical points to the function values at endpoints.

Absolute (Global) Extremes

A function g has an absolute maximum at a point c of its domain if $g(c) \geq g(x)$ for all x in the domain of g.

A function g has an absolute minimum at a point c of its domain if $g(c) \leq g(x)$ for all x in the domain of g.

Figure 5.1 can now be discussed a little more completely. In addition to local extremes, absolute extremes can also be identified. The absolute maximum is

$h(b)$ and the absolute minimum is $h(c)$. It should also be noted that even though they are absolute extremes, both are still listed as local extremes.

EXAMPLE 5.2

Find the absolute maximum and minimum values for $h(x) = x^2 - 2x - 3$ on the closed interval $[0, 3]$.

SOLUTION

Find the critical points by solving $h'(x) = 0$.

$h'(x) = 2x - 2$.

$h'(x) = 0$ at $x = 1$.

$h(1) = -4$

Compare this result to the endpoints of the closed interval.

$h(0) = -3$ and $h(3) = 0$

The absolute minimum value is -4 when $x = 1$.

The absolute maximum value is 0 when $x = 3$.

Not all functions have extremes. Among the infinite assortment of functions, any combination of extremes, local, absolute or no extreme at all—can exist. Consider the graph of $y = x^3$. On its domain, all real numbers, the graph has no local maxima or minima, nor does it have any absolute extremes since $\lim_{x \to \infty} x^3 = \infty$ and $\lim_{x \to -\infty} x^3 = -\infty$. On the other hand, the graph of $f(x) = \sin(x)$ has an infinite number of local maxima and minima, which also happen to be absolute extremes since the range of $\sin(x)$ is $[-1, 1]$!

EXAMPLE 5.3

In Figure 5.5, identify the locations and values of all local and absolute extremes of $g(x)$ on the domain $[-4, 4]$.

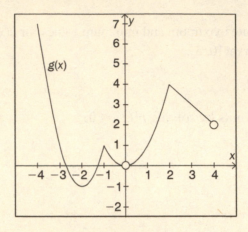

Figure 5.4

SOLUTION

There is a local and absolute maximum of 7 at $x = -4$.

There is a local and absolute minimum of -1 at $x = -2$.

There is a local maximum of 1 at $x = -1$.

There is a local maximum of 4 at $x = 2$.

In Example 5.3, there is a temptation to identify $(0, 0)$ or $(4, 2)$ as local minima, but the discontinuities at those points rule out the possibility. If a point is not in the domain of a function, there is no need to even consider whether an extreme of some kind exists there. For instance, there is no "lowest" point near $(4, 2)$ because whenever some $y_i > 2$ is declared to be the "lowest" y-value, another smaller y_k can be found simply by averaging y_i with 2. The only time a function is actually guaranteed to have extreme values is when it is continuous on a closed interval.

Extreme Value Theorem

> If a function g is continuous on a closed interval [a, b], then
> g will have a maximum and a minimum value somewhere in
> the interval.

EXAMPLE 5.4

Does $f(x) = x^5 - 2x^3 + 3x^2 - 1$ have an absolute maximum on the interval [-2, 1]?

SOLUTION

All polynomials are continuous for all real numbers. Since the interval is closed, the Extreme Value Theorem guarantees an absolute maximum (and minimum) on the interval. Incidentally, the absolute maximum occurs at a critical point in the interior of the domain, but finding the x-value of that critical point by solving $f'(x) = 0$ would require solving a quartic function!

MEAN VALUE THEOREM

Imagine taking a car trip and covering 300 miles in 6 hours. It is a simple matter to determine that the average speed during the trip was 50 miles per hour. Since a car starts from rest, common sense also leads to the conclusion that it would be literally impossible to average 50 miles per hour without at some instant during the trip having the speed of the car be exactly 50 miles per hour. The underlying calculus in this example is based upon an extremely important theorem, which relates instantaneous rate of change to average rate of change during an interval. The actual application of the Mean Value Theorem is usually fairly straightforward, but its greater importance comes from the use of the theorem to develop further significant theorems in the course.

Mean Value Theorem

> If a function f is continuous on the closed interval [a, b] and
> differentiable on the open interval (a, b), then there exists
> at least one point x = c in the open interval (a, b) where
>
> $$f'(c) = \frac{f(b) - f(a)}{b - a}.$$

This theorem says that under the conditions of the hypothesis, the instantaneous rate of change of a function must equal the average rate of change of the function at least once in a given interval. Graphically, recall that the instantaneous rate of change of a function is the slope of the tangent at a point, whereas the average rate of change is the slope of the secant line over an interval. Figure 5.6 provides a graphical look at this principle and actually shows a case with multiple values of x at which the tangent is parallel to the secant. Depending on the function, there may be one, two, or more points where the slope of the tangent will equal the slope of the secant.

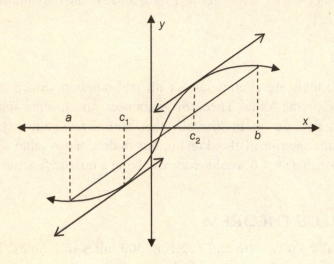

Figure 5.6

EXAMPLE 5.5

Find the value(s) of c that satisfy the Mean Value Theorem for $f(x) = 1 + \sqrt{x}$ on the interval $[0, 9]$.

SOLUTION

$$f'(x) = \left[1 + x^{\left(\frac{1}{2}\right)} \right]$$

$$= \frac{1}{2} x^{\left(-\frac{1}{2}\right)}$$

$$= \frac{1}{2\sqrt{x}}$$

$$m_{\text{sec}} = \frac{f(9) - f(0)}{9 - 0}$$

$$= \frac{4 - 1}{9}$$

$$= \frac{1}{3}$$

We now solve $\dfrac{1}{2\sqrt{c}} = \dfrac{1}{3}$ for c, and Figure 5.7 shows the function, secant, and tangent line.

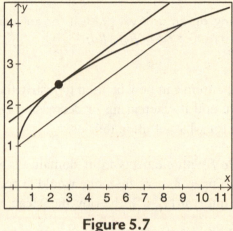

Figure 5.7

$$\frac{1}{2\sqrt{c}} = \frac{1}{3}$$

$$2\sqrt{c} = 3$$

$$\sqrt{c} = \frac{3}{2}$$

$$c = \frac{9}{4}$$

Always check to make sure the values of c lie inside the given interval, especially when more than one value of c is found. Any values of c outside the given interval are not reported in the answer.

INTERVALS OF INCREASING AND DECREASING

Another important application of derivatives is to determine the behavior of functions, particularly where a function is increasing or decreasing. Intuitively, it is not difficult to agree that a function is increasing if whenever x increases, the value of y also increases. Where the domain of a function is continuous, increasing or decreasing behavior can be determined by definition or by derivatives.

Definition of an Increasing/Decreasing Function

> A function f is increasing on an interval $[a,b]$, if for any c and d in the open interval (a,b), $c > d \Rightarrow f(c) > f(d)$.
>
> f is decreasing on an interval $[a,b]$, if for any c and d in the open interval (a,b), $c > d \Rightarrow f(c) < f(d)$.

The Mean Value Theorem can now be used to justify the connection between a function's derivative and its increasing or decreasing behavior. It provides a convincing argument for what intuition tells us.

Consider a function f with elements in its domain $x_2 > x_1$. If f is increasing on the interval containing x_1 and x_2, then one would expect $f(x_2) > f(x_1)$. If f is differentiable on the interval $[x_1, x_2]$, then by the Mean Value Theorem, there exists a value c at which $f'(c) = \dfrac{f(x_2) - f(x_1)}{x_2 - x_1}$. The denominator is positive, since $x_2 > x_1$, and applying the definition of increasing, the numerator is also positive, so for any c, $f'(c)$ must also be positive. A similar argument can also be used for a function decreasing on an interval. These results lead to a simple test for increasing or decreasing behavior of a function on any interval where it is differentiable.

Increasing/Decreasing Derivative Test

> A function is increasing on an interval if $f'(x) > 0$ over the entire interval.
>
> A function is decreasing on an interval if $f'(x) < 0$ over the entire interval.

One way of organizing work is to use a signed number line to test for intervals of increasing or decreasing. Another way is to use a table. The accomplished student may also think graphically about the derivative function to determine where it is positive or negative. What is most important is to find a method that is comfortable and organized for you. The process has essentially two steps: (1) find the critical values, and (2) test the sign of the derivative in the intervals between the critical values. Only one value of the derivative in each interval needs to be tested because the Intermediate Value Theorem guarantees that the derivative could not possibly change signs in an interval without creating an additional zero of the derivative.

EXAMPLE 5.6

Find the intervals of increasing and decreasing for $h(x) = \dfrac{1}{3}x^3 - 3x^2 + 8x - 1$.

SOLUTION

$h'(x) = x^2 - 6x + 8$

Set $h'(x) = 0$ to find critical points.

$x^2 - 6x + 8 = 0$

$(x - 2)(x - 4) = 0$

$x = 2$ or $x = 4$

Test a value of $h'(x)$ in each interval created by $x = 2$ and $x = 4$.

$x < 2$	$2 < x < 4$	$x > 4$
$h'(0) = 8$	$h'(3) = -1$	$h'(5) = 3$

h is increasing where $h' > 0$, in intervals $(-\infty, 2]$ and $[4, \infty)$

h is decreasing where $h' < 0$, in the interval $[2,4]$.

A graphical approach to the values of h' would also have been very efficient. Since $h'(x)$ is a quadratic with a positive leading coefficient, its graph opens upward and has two real zeros. It must be negative between those zeros, and positive elsewhere.

Notice that the endpoints of the intervals were included in the previous example. Textbooks do not agree universally on whether endpoints should be included, so for the purposes of this text, they will be included in intervals of increasing and decreasing.

Another skill to master is the overall connection between the graph of a function, and the graph of its derivative, whether determining characteristics of the function from the derivative graph or vice versa.

EXAMPLE 5.7

Given the graph of $f'(x)$ in Figure 5.8, determine where on the domain $[-4, 4]$ f is increasing or decreasing.

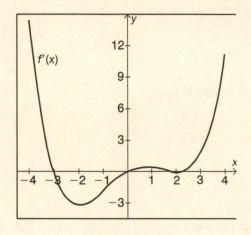

Figure 5.8

SOLUTION

The key to the solution is to realize the y-values must be read from the graph. The y-values are the values of $f'(x)$.

Because $f'(x) > 0$, $f(x)$ is increasing on the intervals $[-4, -3]$ and $[0, 4]$.

Because $f'(x) < 0$, $f(x)$ is decreasing on the interval $[-3, 0]$.

Notice that $f'(2) = 0$, but because $f'(x)$ does not change sign, $f(x)$ continues to increase on the interval $[0, 4]$.

With the ability to determine intervals of increasing and decreasing, a reliable and significant test for local extremes is now available.

First Derivative Test for Local Extremes

> Given that g is a continuous function and c is a critical point, if g' changes from positive to negative at $x = c$, then g has a local maximum at c.
>
> Given that g is a continuous function and c is a critical point, if g' changes from negative to positive at $x = c$, then g has a local minimum at c.

It is necessary for a change of sign to occur. In the case of Figure 5.8, it can be determined that f has a local maximum at $x = -3$ because f' changes from positive to negative at $x = -3$. This means f changes from increasing to decreasing there, so it must have a local maximum. Likewise, f has a local minimum at $x = 0$ because f' changes from negative to positive at $x = 0$. Notice that even though f has a critical point at $x = 2$, there is no maximum or minimum since f' never changes sign at $x = 2$.

EXAMPLE 5.8

Let $h(x)$ be a function such that $h'(x) = (x - 1)[\ln(x)-1]$. Determine the x-values of any local extremes.

SOLUTION

Since $\ln(x)$ is in the problem, the domain is $x > 0$.

Solve $h'(x) = 0$ and look for sign changes in h'.

$(x - 1)\ln(x)-1 = 0$

$(x - 1) = 0$ or $\ln(x) - 1 = 0$

$x = 1$ or $x = e$ since $\ln(e) = 1$.

A number line is used to organize the work and check values of $h'(x)$. It is merely a tool, and not drawn to scale. The actual values tested in $h'(x)$ are not shown, but can be any value in each interval.

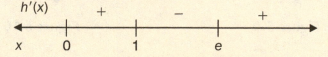

Since h' changes from positive to negative at $x = 1$, h has a local maximum at $x = 1$.

Since h' changes from negative to positive at $x = e$, h has a local minimum at $x = e$.

Always be alert to the required condition of continuity. It is easy to get caught up in using the test and to overlook details such as this. Consider $f(x) = \dfrac{1}{x^2}$. Even though the graph changes from increasing to decreasing at $x = 0$, there is no maximum there because the function is discontinuous at $x = 0$.

5.3   FUNCTIONS AND THEIR SECOND-DERIVATIVE APPLICATIONS

CONCAVITY

As mentioned previously, if the first derivative of a function measures the rate of change of the function values, then the second derivative measures the rate of change of the rate of change. Phrasing it another way, the first derivative represents the slope of a function, so the second derivative measures the rate of change of the slope. The term associated with this property is concavity. If a function is increasing, its slope is positive. If the first-derivative graph, which is a graph of the slope of the original function, is increasing, then its derivative must be positive. Of course, the derivative of the first derivative is the second derivative. So what does a function look like if its slope is increasing? An increasing function with constant slope is simply a line, as shown in Figure 5.9. To have an increasing slope, that line must be "bent upward," as shown in Figure 5.10. This is called "concave up." If the line was "bent downward," it would be called "concave down."

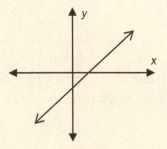

Figure 5.9

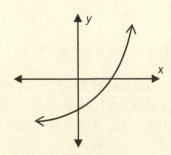

Figure 5.10

Intervals of Concavity

If $g''(x) > 0$ on a given interval, g is said to be concave up on that interval.

If $g''(x) < 0$ on a given interval, g is said to be concave down on that interval.

Essentially, all nonlinear, nonconstant functions can be thought of as consisting of one of four basic shapes made up of the possible combinations of increasing or decreasing, and concave up or concave down. Table 5.1 summarizes the four cases.

Table 5.1

Increasing, concave up	Increasing, concave down	Decreasing, concave up	Decreasing, concave down
$f'(x) > 0$	$f'(x) > 0$	$f'(x) < 0$	$f'(x) < 0$
$f''(x) > 0$	$f''(x) < 0$	$f''(x) > 0$	$f''(x) < 0$

The method for finding intervals of concavity is very similar to the method for finding intervals of increasing or decreasing. The process just uses the second derivative instead of the first derivative, and the points of interest are those values where the second derivative is zero or undefined. Of course, great care must be taken in finding the derivatives, as even the simplest mistake in the derivative expression will result in an incorrect solution.

EXAMPLE 5.9

Find where $f(x) = \sqrt[3]{x}$ is concave up.

SOLUTION

$$f'(x) = \frac{1}{3} x^{\left(-\frac{2}{3}\right)} \text{ and } f''(x) = \frac{-2}{9} x^{\left(-\frac{5}{3}\right)} = \frac{-2}{9\sqrt[3]{x^5}}.$$

$f''(x)$ is undefined at $x = 0$, so examine f'' on each side of 0.

If $x < 0$, $\sqrt[3]{x^5} < 0$, so $f''(x) > 0$.

If $x > 0$, $\sqrt[3]{x^5} > 0$, so $f''(x) < 0$.

Therefore, $f(x) = \sqrt[3]{x}$ is concave up on the interval $(-\infty, 0)$.

Notice that in this case, the right end of the interval is open. The convention of this text will be that intervals of concavity are always open.

EXAMPLE 5.10

Use the graph of $h'(x)$ in Figure 5.11 to determine intervals of concavity for h.

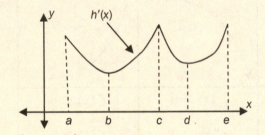

Figure 5.11

SOLUTION

h is concave down on the intervals (a, b) and (c, d) because $h'(x)$ is decreasing, so $h''(x) < 0$ in those intervals.

h is concave up on the intervals (b, c) and (d, e) because $h'(x)$ is increasing, so $h''(x) > 0$ in those intervals.

INFLECTION POINTS

Just as extreme values of a function are significant points of interest, so are extreme values of the first derivative. At these points, the rate of change of a quantity switches from increasing to decreasing or decreasing to increasing.

For example, an economic trend such as the inflation rate may be increasing, but economists are very interested in identifying whether it is increasing at an increasing rate or at a decreasing rate. It is generally good economic news when increasing inflation begins to "slow down." The point at which a function changes concavity is called the inflection point.

Inflection Point

A function f has an inflection point at $x = c$ if $f''(x)$ changes signs at $x = c$, and if a tangent line can be drawn at $(c, f(c))$.

In order for f'' to change signs at $x = c$, either $f''(c) = 0$ or f'' must be undefined. Notice the extension to thinking of the first-derivative function as its own entity, and f'' as its first derivative. Seeking inflection points begins with finding critical points of $f'(x)$! A key subtlety lies in the additional connection to the original function, f. It is possible for f'' to change signs at a point of f without having an inflection point on f. It is also possible to have an inflection point on a function where the first and second derivatives of the function are undefined.

Figure 5.12 shows a change in concavity without an inflection point. Figure 5.13 shows an inflection point existing at $(1, 1)$ even though the first

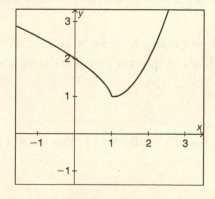

Figure 5.12

derivative is undefined. The key is that there is a vertical tangent at $(1, 1)$. Informally, the transition from concave up to concave down or down to up must be "smooth," as in Figure 5.13.

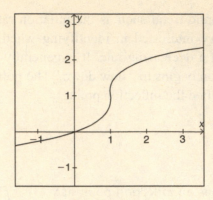

Figure 5.13

EXAMPLE 5.11

If $f(x) = x^3 - 4x^2 + 5$, find the x-values of each inflection point on f.

SOLUTION

$f'(x) = 3x^2 - 8x$, and $f''(x) = 6x - 8$.

Solving $f''(x) = 0$ gives $6x - 8 = 0$, so $x = \dfrac{4}{3}$.

Since f'' is linear with positive slope, it changes from negative to zero to positive at $x = \dfrac{4}{3}$, so $f(x) = x^3 - 4x^2 + 5$ has an inflection point at $x = \dfrac{4}{3}$.

Be aware that not all problems are quite so simple as the last. The complexity of the second derivative can quickly increase the difficulty of a problem.

EXAMPLE 5.12

Find the coordinates of all inflection points on $h(x) = \dfrac{1}{x^2 + 1}$.

SOLUTION

Think of h as $h(x) = (x^2 + 1)^{-1}$ and use the power rule with the chain rule.

$h'(x) = -1(x^2 + 1)^{-2} \cdot 2x$

$\qquad = -2x(x^2 + 1)^{-2}$

Now the product rule is necessary to find h''.

$$h''(x) = -2x[-2(x^2+1)^{-3}] \cdot 2x + (x^2+1)^{-2} \cdot (-2)$$

$$= \frac{8x^2}{(x^2+1)^3} + \frac{-2}{(x^2+1)^2}$$

Solve $h''(x) = 0$ by adding the fractions and setting the numerator equal to 0. The denominator is never equal to 0, so $h''(x)$ is defined over all real numbers.

$$\frac{8x^2}{(x^2+1)^3} + \frac{-2}{(x^2+1)^2} = 0$$

$$\frac{8x^2}{(x^2+1)^3} + \frac{-2(x^2+1)}{(x^2+1)^3} = 0$$

$$\frac{8x^2 - 2x^2 - 2}{(x^2+1)^3} = 0$$

$$\frac{6x^2 - 2}{(x^2+1)^3} = 0$$

For $h''(x) = 0$, $6x^2 - 2 = 0$, so $x = \pm\sqrt{\frac{1}{3}}$.

Now check concavity in the intervals determined by x. Use its simplest form $h''(x) = \frac{6x^2-2}{(x^2+1)^3}$.

$x < -\sqrt{\dfrac{1}{3}}$	$-\sqrt{\dfrac{1}{3}} < x < \sqrt{\dfrac{1}{3}}$	$x > \sqrt{\dfrac{1}{3}}$
$h''(-1) = \dfrac{4}{8}$	$h''(0) = -2$	$h''(1) = \dfrac{4}{8}$
Concave up	Concave down	Concave up

$\left(-\sqrt{\dfrac{1}{3}}, h\left(-\sqrt{\dfrac{1}{3}}\right)\right) = \left(-\sqrt{\dfrac{1}{3}}, \dfrac{3}{4}\right)$ and $\left(\sqrt{\dfrac{1}{3}}, h\left(\sqrt{\dfrac{1}{3}}\right)\right) = \left(\sqrt{\dfrac{1}{3}}, \dfrac{3}{4}\right)$ are inflection points of h.

Clearly, more attention to detail is required on certain problems than on others, but the concepts and relationships between a function and its multiple derivatives are just as important to master. The following example involves no analytic work, but does require a solid conceptual understanding.

EXAMPLE 5.13

Let h be differentiable over its entire domain. Use the graph of $h'(x)$ in Figure 5.14 to determine the x-coordinates of each inflection point on h.

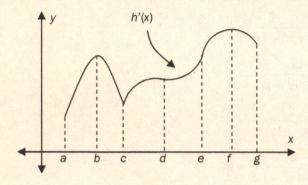

Figure 5.14

SOLUTION

The inflection points occur at the extremes of $h'(x)$, $x = b$, $x = c$, and $x = f$.

From a to b, $h'(x)$ is increasing, so $h''(x) > 0$, meaning $h(x)$ is concave up. From b to c, $h'(x)$ is decreasing, so $h''(x) < 0$, meaning $h(x)$ is concave down. Since h is differentiable and changes concavity at $x = b$, there is an inflection point there. Similar logic is used for $x = c$ and $x = f$. Even though there is a corner on h' at $x = c$, it does not prevent h from having an inflection point there. As long as h is differentiable, and its derivative has an extreme value on the interior of the domain, an inflection point will exist.

Besides being used to determine intervals of concavity and inflection points, the second derivative can also be applied efficiently to justify the existence of a local extreme value on a function. Recall that the first-derivative test for maxima and minima required identifying critical points and checking whether the first derivative changed signs on either side of the critical point. This requires knowledge about how the function is behaving around the critical point. The second-derivative test is a local test applied right at the critical point.

Second Derivative Test for Local Extremes

Let c be a point on the interior of the domain of a function, f.

If $f'(c) = 0$ and $f''(c) > 0$, then f has a local minimum at $x = c$.

If $f'(c) = 0$ and $f''(c) < 0$, then f has a local maximum at $x = c$.

If $f'(c) = 0$, f has a horizontal tangent at $x = c$. But if $f'(c) > 0$, then f is concave up, so the graph must lie above the horizontal tangent at c. Therefore, f must have a local minimum at $x = c$. Similar logic can be applied for a local maximum.

EXAMPLE 5.14

Use the second-derivative test to determine the x-coordinates of the local extreme values on $f(x) = x^3 + 3x^2 - 9x + 2$.

SOLUTION

$f'(x) = 3x^2 + 6x - 9$, and $f''(x) = 6x + 6$.

Solving $f'(x) = 0$ gives $3(x^2 + 2x - 3)$ or $3(x + 3)(x - 1) = 0$.

So the critical values are $x = -3$ and $x = 1$.

$f''(-3) = -12$, and $f''(1) = 12$, so $f''(-3) < 0$ and $f''(1) > 0$.

By the second-derivative test, f has a local maximum at $x = -3$ and a local minimum at $x = 1$.

CURVE SKETCHING

Mastering the relationships between a function and its first and second derivatives takes a good amount of practice. One way to test your conceptual understanding of these relationships is to work graphically with the three functions. The goal is to be able to synthesize information in such a way as to move seamlessly between a function and its derivatives. Given information about first and second derivatives, you ought to be able to make a rough sketch of the function. Given a function, you ought to be able to sketch the graph of its derivative.

EXAMPLE 5.15

Given the piecewise function $g(x)$ shown in Figure 5.15, sketch its derivative $g'(x)$.

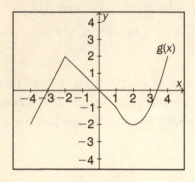

Figure 5.15

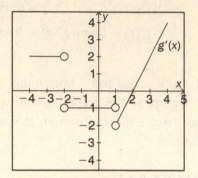

Figure 5.16

SOLUTION

The derivative graph is shown in Figure 5.16. On the interval $[-4, -2]$, g is linear with a constant slope of 2. For $[-2, 1]$, g is linear with a constant slope of -1. For $[1, 2]$, g is decreasing to a local minimum, where $g'(x) = 0$. On the interval $[2, 4]$, $g'(x)$ is positive and getting more positive as x increases. At $x = -2$ and $x = 1$, g has corners so $g'(x)$ does not exist. There are no open circles at the endpoints of $g'(x)$ because the one-sided derivatives at each endpoint of g are defined.

EXAMPLE 5.16

Sketch a graph of $f(x)$ on the domain $[-2, 4]$, by using the following information:

$$f(-2) = 3, f(-1) = 1, f(2) = 4$$

$f'(x) < 0$ for $-2 < x < -1$ and $2 < x < 4$	$f'(x) > 0$ for $-1 < x < 2$
$f'(-1)$ does not exist	$f'(2) = 0$
$f''(x) < 0$ for $-2 < x < -1$ and $1 < x < 4$	$f''(x) > 0$ for $-1 < x < 1$

SOLUTION

Place points for the three ordered pairs first. Lightly draw segments for the intervals of increasing and decreasing by using the $f'(x)$ information. Use concavity information from $f''(x)$ to curve the lightly drawn segments. The graph of $f(x)$ drawn from this information is shown in Figure 5.17.

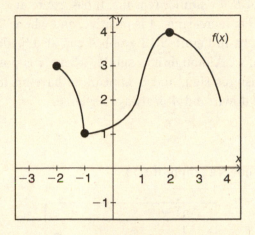

Figure 5.17

EXAMPLE 5.17

Identify the three graphs in Figure 5.18 as g, g' and g''.

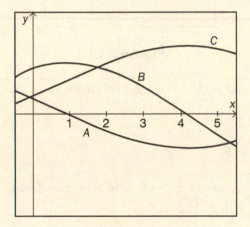

Figure 5.18

SOLUTION

Graph C is g.

Graph B is g'.

Graph A is g''.

B is the derivative of C because it has a zero where C has a maximum. Also, when C is increasing, B is positive, and when C is decreasing, B is negative. A is the derivative of B for similar reasons. It has a zero at $x = 1$, where B has a maximum. Where B is increasing, A is positive, and where B is decreasing, A is negative. Interestingly, in Figure 5.18 graph A can also be determined to be the second derivative of C. Although it is subtle, where A is positive, C is concave up. At $x = 1$, A changes sign, and C appears to have an inflection point. For $x > 1$, C is concave down, and A is always negative.

EXAMPLE 5.18

Figure 5.19 shows a graph of $f(x)$.

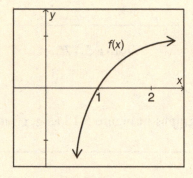

Figure 5.19

Order $f(1), f'(1),$ and $f''(1)$ from least to greatest.

SOLUTION

$$f''(1) < f(1) < f'(1)$$

Check the three values at $x = 1$. f is concave down, so $f''(1) < 0$. f is increasing, so $f'(1) > 0$. $f(1) = 0$.

5.4 LINEAR PARTICLE MOTION

Linear particle motion is another derivative application. It also appears frequently in physics problems. The task involves making connections between the position of a particle moving along the x-axis, its velocity, and its acceleration. The functions can be defined by using any variables, but frequently used notation for position is $x(t)$ or $s(t)$, the x (or s) position as a function of time. Naturally, velocity is denoted by $v(t)$, and its derivative, acceleration, by $a(t)$. As a particle moves to the right, the x-coordinate is increasing as time is increasing, so the ratio for average velocity, $\dfrac{\Delta x}{\Delta t}$, is positive. As the time interval shortens, the instantaneous velocity, $\lim\limits_{\Delta t \to 0} \dfrac{\Delta x}{\Delta t} = \dfrac{dx}{dt}$, is also defined as positive. Similarly, motion to the left defines negative velocity. Often, over a time interval, a particle may move right and left. Although this motion has little impact on instantaneous velocity or acceleration, it has significant implications for average velocity. Average velocity depends on displacement, and displacement is defined simply as final position minus initial position.

KEY CONCEPTS OF HORIZONTAL PARTICLE MOTION

Displacement

Let $x(t)$ be the position of a particle moving along the x-axis. Position is a function of time, t. On a given time interval $[a, b]$ the particle's, displacement $= \Delta x = x(b) - x(a)$.

Average Velocity

If $x(t)$ is the position of a particle moving along the x-axis, on a given time interval $[a, b]$, its average velocity
$= \dfrac{\Delta x}{\Delta t} = \dfrac{x(b) - x(a)}{b - a}$.

Instantaneous Velocity

If $x(t)$ is the position of a particle moving along the x-axis,

instantaneous velocity $= \lim\limits_{\Delta t \to 0} \dfrac{\Delta x}{\Delta t} = \dfrac{dx}{dt}$.

The instantaneous velocity of the particle measured at any moment $t = c$ is $v(c) = x'(c)$.

Speed

Speed is the absolute value of velocity. It does not take into account direction of motion.

Speed at any moment $t = c$ is $|v(c)|$

Acceleration

If $x(t)$ is the position of a particle moving along the x-axis, its

$$\text{acceleration} = \lim_{\Delta t \to 0} = \frac{\Delta v}{\Delta t} = \frac{dv}{dt}.$$

The acceleration of the particle measured at any moment $t = c$, is $a(c) = v'(c) = x''(c)$.

EXAMPLE 5.19

The position of a particle moving along the x-axis is defined by $x(t) = t \cdot \sin(2t)$. Find the displacement of the particle on the time interval, $t = 0$ to $t = \dfrac{3\pi}{2}$ seconds.

SOLUTION

$$x\left(\frac{3\pi}{2}\right) - x(0) = \frac{3\pi}{2}\sin(3\pi) - 0\sin(0)$$
$$= 0$$

The particle has oscillated during the interval, but has returned to its starting position when $t = \dfrac{3\pi}{2}$ seconds, so there is no displacement.

EXAMPLE 5.20

The position of a particle moving along the x-axis is defined by $x(t) = \ln(t^2 + 1)$ for $t \geq 0$. If distance is measured in feet and time is in seconds, find the average acceleration on the time interval $[1, 3]$ seconds.

SOLUTION

If average velocity is change in position over change in time, by extension, average acceleration is change in velocity over change in time.

$$\text{Velocity} = v(t) = x'(t) = \frac{2t}{t^2 + 1}$$

$$\text{Average acceleration} = \frac{v(3) - v(1)}{3 - 1} = \frac{\frac{6}{10} - \frac{2}{2}}{2} = \frac{-1}{5} \text{ ft/sec}^2$$

EXAMPLE 5.21

The position of a particle moving along the x-axis is defined by

$$x(t) = \frac{1}{2} t^4 - 4t^3 + 8t^2 + 1.$$

For $t \geq 0$, find when the particle is moving right.

SOLUTION

The particle is moving right when its velocity is positive. Find when the velocity is 0 and test the intervals between those times.

$v(t) = x'(t) = 2t^3 - 12t^2 + 16t$

$2t^3 - 12t^2 + 16t = 0$

$2t(t^2 - 6t + 8) = 0$

$2t(t - 2)(t - 4) = 0 \Rightarrow t = 0$ or $t = 2$ or $t = 4$.

$0 < t < 2$	$2 < t < 4$	$t > 4$
$v(1) = 6$	$v(3) = -6$	$v(5) = 30$

The particle is moving right in the time intervals $0 < t < 2$ and $t > 4$.

EXAMPLE 5.22

A particle is moving along the x-axis. Figure 5.20 shows its velocity graph. On the interval [0, 6], when is the speed of the particle greatest?

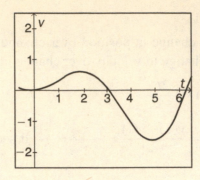

Figure 5.20

SOLUTION

Since speed is the absolute value of velocity, speed is greatest at about $t = 5$ seconds, not 2 seconds.

5.5   OPTIMIZATION

Optimization is one of the most interesting applications of derivatives. It applies maximizing and minimizing to practical problems, and it involves a wide diversity of problems. Fortunately, there is an organized approach that can be taken to help solve these types of problems with a high degree of success.

Practical Steps to Solving Optimization Problems

1. Carefully read the problem and make a labeled drawing.

2. List parameters that will not change throughout the problem.

3. Identify which quantity is to be optimized, and write it as a function of the variables in the labeled drawing.

4. If what is to be optimized is a function of more than one variable, substitution should be made by using other fixed relationships in the problem.

5. When the function to be optimized is a function of one variable, use differentiation to find the extreme values in the practical domain.

A special word of caution is required here. When working on problems in the Cartesian coordinate plane, never redefine x or y to be anything other than their intrinsic definitions—the directed horizontal and vertical distances from the origin. Redefining x or y creates a risk of changing the equation of any functions given in the original problem.

EXAMPLE 5.23

Find the dimensions of a rectangle with the greatest area that can be placed with one corner at the origin and the opposite corner on the line $y = 6 - 2x$.

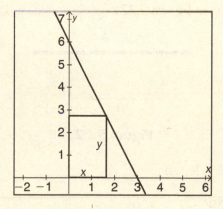

Figure 5.21

SOLUTION

Sketch the problem (see Figure 5.21). From the sketch, the practical domain is shown to be [0, 3].

Let x = width and y = height.

Then area is $A(x) = x \cdot y$, but since $y = 6 - 2x$, by substitution $A(x) = x(6 - 2x)$.

$A(x) = 6x - 2x^2 \quad A'(x) = 6 - 4x$

$6 - 4x = 0 \Rightarrow x = \dfrac{3}{2}$

$A''(x) = -4$. By the second-derivative test, $x = \dfrac{3}{2}$ creates a local maximum on $A(x)$. At the endpoints of the domain, the area of the rectangle is 0, so the local maximum is also the absolute maximum.

The dimensions of the rectangle with the greatest area are width $= \dfrac{3}{2}$ and height $= 6 - 2 \cdot \dfrac{3}{2} = 3$.

EXAMPLE 5.24

Three sides of a rectangular region adjacent to a barn need to be fenced. What is the largest area that can be fenced in with 120 feet of fence?

SOLUTION

Sketch the problem (see Figure 5.22).

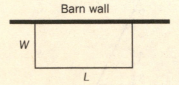

Barn wall

Figure 5.22

Let W = width and L = length.

Area $= W \cdot L$, and $2W + L = 120$.

Isolate L to get $L = 120 - 2W$, so

$A(W) = W(120 - 2W) = 120W - 2W^2$.

$A'(W) = 120 - 4W$

$120 - 4W = 0 \Rightarrow W = 30$

If $W = 30$, then $L = 120 - 2 \cdot 30 = 60$

Area $= 30 \cdot 60 = 1800$ sq. ft.

EXAMPLE 5.25

Find the radius of the cylinder of greatest volume that can be inscribed inside an upright cone with base radius of 4 inches and height of 12 inches.

SOLUTION

Sketch the problem (see Figure 5.23).

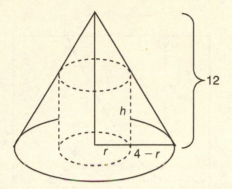

Figure 5.23

The volume of a cylinder is $V = \pi r^2 h$.

Since V is a function of two variables, a substitution must take place to eliminate one of the variables. Often, when triangles are involved in a problem, either the Pythagorean Theorem or similar triangles may come into play. In this problem, similar triangles give the proportion, $\dfrac{4}{12} = \dfrac{4-r}{h}$.

Solving for h, $h = 12 - 3r$.

Substitute this expression for h into the volume function to get $V = \pi r^2 (12 - 3r)$.

$$V = \pi(12r^2 - 3r^3)$$

$$V'(r) = \pi(24r - 9r^2)$$

$$\pi(24r - 9r^2) = 0$$

$$\pi \cdot 3r(8 - 3r) = 0 \Rightarrow r = 0 \text{ or } r = \frac{8}{3}$$

$V''(r) = \pi(24 - 18r)$ so $V''\left(\dfrac{8}{3}\right) < 0$, which makes $r = \dfrac{8}{3}$ the radius that maximizes the volume of the cylinder.

EXAMPLE 5.26

Which point on the graph of $f(x) = x^{\left(\frac{3}{2}\right)}$ shown in Figure 5.24 is closest to the point (2.5, 0)?

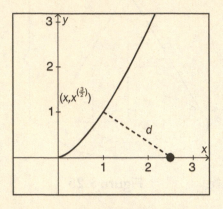

Figure 5.24

SOLUTION

The objective is to minimize the distance between (2.5, 0) and any point on $f(x) = x^{\left(\frac{3}{2}\right)}$.

It is actually easier to minimize the square of the distance and still find the proper point.

Any point on the curve can be labeled $\left(x, x^{\left(\frac{3}{2}\right)}\right)$.

The Distance Formula gives

$$d = \sqrt{(x-2.5)^2 + \left(x^{\left(\frac{3}{2}\right)} - 0^2\right)^2}, \text{ so } d^2 = \left(x - 2.5\right)^2 + (x^{\left(\frac{3}{2}\right)} - 0)^2.$$

Let $D = d^2$, so $D(x) = (x - 2.5)^2 + x^3$.

Then $D'(x) = 2(x - 2.5) + 3x^2$.

Solve for x when $D'(x) = 0$ to find the extremes.

$3x^2 + 2x - 5 = 0$

$$(3x+5)(x-1) = 0 \Rightarrow x = 1 \text{ or } x = \frac{-5}{3}$$

$x = \dfrac{-5}{3}$ is not in the domain of $f(x)$, so it is disregarded.

$D''(x) = 6x + 2$, and $D''(1) > 0$, so $x = 1$ determines a local minimum.

The point on $f = x^{\left(\frac{3}{2}\right)}$ closest to (2.5, 0) is thus (1, 1).

5.6   RELATED RATES

Whenever one or more objects are in motion, their positions relative to one another or any objects around them can be considered functions of time. For example, when two jets at different altitudes fly on skew paths, the distance between the planes is constantly changing and is related to their rates and positions. If two people are walking straight toward each other, and one person is walking at 4 miles per hour while the other is walking at 3 miles per hour, the distance between the people is decreasing at 7 miles per hour. Keeping this in mind, when working with related rates, the variable of differentiation is t, time.

The process begins similarly to that of optimization. It is important to carefully define variables and the function or functions that relate the quantities in a problem. It is also advantageous to work with single variable functions when possible.

Several key differences from optimization are of note as well. First, the variable of differentiation, t, is implicit, but appears in rates through the use of the chain rule. Second, rates of change are represented by using Leibniz notation, such as $\dfrac{dr}{dt}$. Finally, for related rates, the derivative expressions are not set equal to zero and solved. When done correctly, a related-rates equation created by differentiation contains one unknown, which is to be determined. When done incorrectly, the most common student mistake is substituting values that could change into a function expression prior to differentiation. It is important to always be aware of what is changing and what is constant in a problem. Therefore, it is a good idea to substitute known values only after differentiation has occurred.

Practical Steps to Solving Related Rates Problems

1. Carefully read the problem and make a labeled drawing.

2. Write down all rates and known values.

3. Write down all functions that relate the variables in the labeled drawing.

4. Determine which function needs to be differentiated to introduce the desired rate into the problem.

5. If what is to be differentiated is a function of more than one variable, substitute using other fixed relationships in the problem so there is only one unknown.

6. Differentiate with respect to t, time.

7. Substitute known values and solve for the desired rate.

EXAMPLE 5.27

If the sides of a square are functions of time, differentiate $A = s^2$ with respect to t.

SOLUTION

One solution

$$\frac{d}{dt}(A) = \frac{d}{dt}(s^2)$$

$$\frac{dA}{dt} = 2s^{2-1}\frac{d}{dt}(s)$$

$$\frac{dA}{dt} = 2s\frac{ds}{dt}$$

Alternate solution

$$\frac{dA}{ds} = 2s$$

$$\frac{dA}{dt}\frac{ds}{dt} = 2s\frac{ds}{dt}$$

$$\frac{dA}{dt} = 2s\frac{ds}{dt}$$

EXAMPLE 5.28

Air is being blown into a spherical balloon at a rate of 4 cubic inches per minute. How fast is the radius changing at the instant when the radius is 3 inches?

SOLUTION

Known values: $\dfrac{dV}{dt} = 4$ and $r = 3$ (Do not substitute until after differentiating!)

Volume of a sphere: $V = \dfrac{4}{3}\pi r^3$

$$\frac{d}{dt}(V) = \frac{d}{dt}\left(\frac{4}{3}\pi r^3\right) \Rightarrow \frac{dV}{dt} = \frac{4}{3}\pi \cdot 3r^2 \frac{dr}{dt} = 4\pi r^2 \frac{dr}{dt}$$

Much like the case of implicit differentiation, $\dfrac{dr}{dt}$ is the derivative of the base of r differentiated with respect to t. Now substitute known values.

$$4 = 4\pi \cdot 3^2 \frac{dr}{dt} \Rightarrow \frac{dr}{dt} = \frac{1}{9\pi} \text{ inches per minute.}$$

EXAMPLE 5.29

An inverted cone with base radius 2 inches and height 8 inches is being filled with water at a rate of 1 cubic inch per minute. How fast is the radius changing at the instant the depth of the water is 4 inches?

SOLUTION

Make a labeled drawing (Figure 5.25).

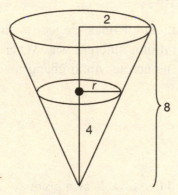

Figure 5.25

Known values: $\dfrac{dV}{dt} = 1$ and $h = 4$.

Volume of a cone: $V = \dfrac{1}{3}\pi r^2 h$

Notice that volume is a function of two variables. Use similar triangles to substitute for h in terms of r. At any moment, the radius and height of the water are in the same ratio as the radius and height of the cone.

Thus, $\dfrac{h}{r} = \dfrac{8}{2}$ and $h = 4r$.

$$V = \frac{1}{3}\pi r^2 \cdot 4r = \frac{4}{3}\pi r^3$$

$$\frac{dV}{dt} = \frac{4}{3}\pi \cdot 3r^2 \frac{d}{dt}(r) = 4\pi r^2 \frac{dr}{dt}$$

Since $h = 4r$, when $h = 4$, $r = 1$.

Substitute known values into $\dfrac{dV}{dt} = 4\pi r^2 \dfrac{dr}{dt}$.

$$1 = 4\pi \cdot 1^2 \frac{dr}{dt} \implies \frac{dr}{dt} = \frac{1}{4\pi} \text{ inches per minute.}$$

5.7 LINEARIZATION

Earlier, you wrote equations of lines tangent to functions. Here, we revisit the topic briefly to look at it in a different way. When a linearization of a non-linear function is written, numerous small changes occur that can and should be examined.

Linearization

> If a function g is differentiable at a point $x = a$, the linearization is $L(x) = g(a) + g'(a)(x - a)$.

Even though the given equation looks somewhat complex, it is only a rearrangement of the point-slope form of the equation of the tangent line to g at the point $(a, g(a))$. The emphasis of this second look, however, is on estimation and the error involved.

Part of the notation introduces the idea of differentials. Any change in x can be written by using the symbol Δx, but when that change becomes infinitely small, the differential dx is used. Likewise, Δy may be replaced with the differential dy, and when the derivative is defined for a function $g(x)$, $dy = g'(x)dx$. Another convention used in many applications is for Δy to represent actual change in a function value, and for dy to represent approximate change. This is because Δy is measuring change in actual function values, whereas dy is measuring change by using the approximating tangent line.

Figure 5.26 provides a graphical look at the various elements involved in linear approximation and error, including differentials.

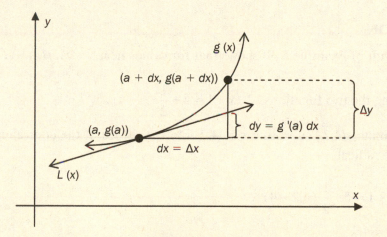

Figure 5.26

The error in approximating a function is the length of the segment running vertically from the line $L(x)$ to the point $(a + dx, g(a + dx))$. By examining Figure 5.26 you can see that error is simply $|\Delta y - dy|$.

EXAMPLE 5.30

Write the linearization of $f(x) = 2\sqrt{x} + 1$ at $x = 4$.

SOLUTION

$f(4) = 2\sqrt{4} + 1 = 5$, so the tangent is being drawn at the point (4, 5).

$f'(x) = \dfrac{1}{\sqrt{x}}$ and $f'(4) = \dfrac{1}{2}$, so the slope of the tangent is $\dfrac{1}{2}$.

The linearization is $y - 5 = \dfrac{1}{2}(x - 4)$ or, using linearization notation,

$L(x) = 5 + \dfrac{1}{2}(x - 4)$.

EXAMPLE 5.31

Use the linearization found in Example 5.30 to estimate $\sqrt{4.2}$.

SOLUTION

The result of Example 5.30 shows that for values near $x = 4$, $f(x) \approx L(x)$.

Equating the two functions, $2\sqrt{x} + 1 \approx 5 + \dfrac{1}{2}(x - 4)$.

To estimate $\sqrt{4.2}$, substitute $x = 4.2$ into both sides of the equivalence and isolate the radical.

$2\sqrt{4.2} + 1 \approx 5 + \dfrac{1}{2}(4.2 - 4)$

$2\sqrt{4.2} \approx 4.1$

$\sqrt{4.2} \approx 2.05$

It is interesting to note that $\sqrt{4.2}$ by calculator gives 2.04939...

EXAMPLE 5.32

Use differentials to estimate the change in the volume of a cube as the length of one edge changes from 4 to 4.1 inches. Compare the result to the actual change.

SOLUTION

For a cube with edges of length s, volume $V = s^3$.

$V'(s) = 3s^2$, so $dV = 3s^2 ds$.

$dV = 3(4^2)(0.1) = 4.8$, or the approximate change is 4.8 cubic inches.

The actual change is $\Delta V = 4.1^3 - 4^3 = 4.921$ cubic inches. The error is 0.121 cubic inches.

5.8   L'HÔPITAL'S RULE

With a stronger understanding of the linearization of differentiable functions, we now introduce L'Hôpital's Rule for evaluating limits. From time to time, the evaluation of limits leads to a quandary when both the numerator and denominator of a fraction appear to head toward zero because $\dfrac{0}{0}$ is undefined! At times, the numerator and denominator may also both grow without bound to a form such as $\dfrac{\infty}{\infty}$. Often, resourceful algebraic methods help to resolve the issue, but sometimes they don't. Expressions such as these are indeterminate forms and may often be dealt with by using L'Hôpital's Rule. A complete course would address a wide variety of indeterminate forms, but this text examines only the most common forms.

L'Hôpital's Rule

If functions f and g are both differentiable at $x = a$, and $f(a) = g(a) = 0$, then $\displaystyle\lim_{x \to a} \frac{f(x)}{g(x)} = \lim_{x \to a} \frac{f'(x)}{g'(x)}$ provided $g'(a) \neq 0$.

Remarkably, what L'Hôpital's Rule is saying is that given the conditions of the hypothesis, the ratio of the function values is equal to the ratio of derivatives. Figure 5.27 and the explanation that follows provide intuitive support for the rule.

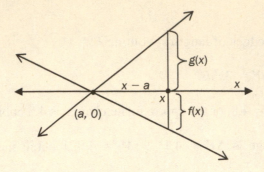

Figure 5.27

If the functions are differentiable at $x = a$, then they will have local linearity. Figure 5.27 is a "microscopic" look at the functions near $x = a$. The slope of each linearized function is $\dfrac{f(x)}{x-a}$ and $\dfrac{g(x)}{x-a}$. But the slope of the functions near a are also $f'(a)$ and $g'(a)$. Therefore, for x values very close to a, $\dfrac{f'(x)}{g'(x)} = \dfrac{\frac{f(x)}{x-a}}{\frac{g(x)}{x-a}} = \dfrac{f(x)}{g(x)}$. As x gets ever closer to a, $\displaystyle\lim_{x \to a} \dfrac{f(x)}{g(x)} = \lim_{x \to a} \dfrac{f'(x)}{g'(x)}$.

To apply L'Hôpital's Rule, the ratio being evaluated must be an indeterminate form, or be able to be rewritten as one. An alternate version of L'Hôpital's Rule applies to ratios for which the numerator and denominator both approach $\pm\infty$, as presented next.

L'Hôpital's Rule

If functions f and g are both differentiable at $x = a$, and f and g approach infinity as x approaches a, then $\displaystyle\lim_{x \to a} \dfrac{f(x)}{g(x)} = \lim_{x \to a} \dfrac{f'(x)}{g'(x)}$ provided $g'(a) \neq 0$.

EXAMPLE 5.33

Evaluate $\displaystyle\lim_{x \to 0} \dfrac{\cos(2x) + 3x - 1}{x^2 - x}$.

SOLUTION

At $x = 0$, the limit is indeterminate, $\dfrac{0}{0}$. Differentiate the numerator and denominator and examine the limit of the ratio of the derivatives.

$$\lim_{x \to 0} \frac{\cos(2x) + 3x - 1}{x^2 - x} = \lim_{x \to 0} \frac{-2\sin(2x) + 3}{2x - 1}$$

$$= \frac{0 + 3}{0 - 1} = -3$$

EXAMPLE 5.34

Evaluate $\displaystyle\lim_{x \to \infty} \frac{3x^2 + 4}{5x^2 - 9}$.

SOLUTION

$\displaystyle\lim_{x \to \infty} \frac{3x^2 + 4}{5x^2 - 9}$ approaches $\dfrac{\infty}{\infty}$, so L'Hôpital's Rule can be applied.

$$\lim_{x \to \infty} \frac{3x^2 + 4}{5x^2 - 9} = \lim_{x \to \infty} \frac{6x}{10x}$$

$$= \frac{3}{5}$$

This result should be no surprise because this type of limit was discussed in Chapter 2. Since the degree of the numerator and denominator are equal, the limit is the ratio of the leading coefficients.

EXAMPLE 5.35

Evaluate $\displaystyle\lim_{x \to 2} \frac{x^3 - 3x^2 + 4}{x^2 - 4}$.

SOLUTION

The form is $\dfrac{0}{0}$, so apply L'Hôpital's Rule.

$$\lim_{x \to 2} \frac{x^3 - 3x^2 + 4}{x^2 - 4} = \lim_{x \to 2} \frac{3x^2 - 6x}{2x}$$

At this point there is a great temptation to differentiate again, but the current limit is not an indeterminate form, so L'Hôpital's Rule does not apply. Make a habit of checking for an indeterminate form after each step.

$$\lim_{x \to 2} \frac{3x^2 - 6x}{2x} = \frac{3(2)^2 - 6(2)}{2(2)}$$

$$= \frac{0}{4}$$

$$= 0$$

5.9 EXERCISES

1. Find each local extreme value for $f(x) = -2x^3 - x^2 + 8x - 5$.

2. The graph of $h'(x)$ is given in Figure 5.28. Determine the x-coordinates of each local extreme value. Identify each as a local maximum or minimum.

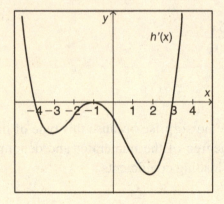

Figure 5.28

3. Find the absolute maximum and minimum values of $g(x) = \ln(x) + \frac{1}{x}$ on the interval $\frac{1}{e} \le x \le e$.

4. Determine the intervals of concavity for $f(x) = \tan^{-1}(x + 1)$

Table 5.2

x	1	2	3	4	5	6
$f(x)$	2.3	4.5	6.7	6.3	4.5	1.9

5. Table 5.2 lists ordered pairs for $f(x)$. $f(x)$ is differentiable for all real numbers. Explain why f must have at least one horizontal tangent in the interval $(2, 5)$.

6. Find each value of c that satisfies the Mean Value Theorem for $g(x) = x^{\left(\frac{2}{3}\right)}$ on the interval $[0, 8]$.

7. Given that h is a continuous function for all real numbers, and that $h'(x) = \dfrac{x^2(x-8)(x+1)}{x-2}$, determine the intervals on which h is increasing and decreasing.

8. If $f'(x) = x^2(x - 3)$, find the x-coordinate of each point of inflection on the graph of f.

9. The graph of $h'(x)$ is shown in Figure 5.29. Use the graph to estimate the x-coordinates of any local extreme values for $h(x)$.

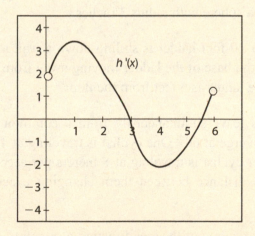

Figure 5.29

10. Use Figure 5.29 to determine any interval where $h(x)$ is concave up.

11. If $h(5) = 3$, use Figure 5.29 to write the linearization of $h(x)$ at $x = 5$.

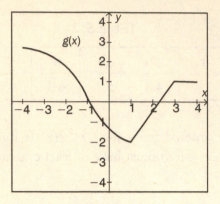

Figure 5.30

12. The graph of $g(x)$ is shown in Figure 5.30. Sketch a graph of $g'(x)$ on the domain $[-4, 4]$.

13. The position of a particle moving along the x-axis is defined by $x(t) = \sin\left(\dfrac{\pi}{3}t\right)$. In the time interval $[0, 5]$ seconds, when is the particle stationary?

14. Find the height and radius of the right cylindrical can with volume 16π cubic inches that uses the least amount of material.

15. What is the volume of the largest square-based pyramid that can be inscribed in a sphere with radius 3 inches?

16. The top of a 10-foot ladder is sliding down a wall at 3 feet per second. How fast is the base of the ladder moving away from the wall the instant the top of the ladder is 6 feet from the floor?

17. Two cyclists leave simultaneously from a common spot and travel on paths that diverge at 60°. One cyclist is traveling at 4 meters per second, and the other cyclist is traveling at 8 meters per second. How fast is the straight line distance between them changing exactly 5 seconds after they leave?

18. Use the linearization of the area function at $r = 6$ to estimate the change in the area of a circle as its radius increases from 6 to 6.2 inches.

19. Find the linearization of $g(x) = \sin\left(\dfrac{x}{2}\right)$ at $x = 0$ and use it to estimate $\sin\left(\dfrac{\pi}{6}\right)$.

20. Evaluate $\displaystyle\lim_{x\to 1}\frac{\cos(\pi x) + x}{x^2 - 1}$.

21. Evaluate $\displaystyle\lim_{x\to\infty}\frac{\ln(3x)}{\sqrt{x}}$.

5.10 SOLUTIONS TO EXERCISES

1. Local minimum at $x = \dfrac{-4}{3}$ and local maximum at $x = 1$.

   $f(x) = -2x^3 - x^2 + 8x - 5$, so solve $f'(x) = -6x^2 - 2x + 8 = 0$ to find extremes.

   $$-2(3x^2 + x - 4) = 0$$
   $$-2(3x + 4)(x - 1) = 0$$

   The zeros of $f'(x)$ are $x = \dfrac{-4}{3}$ and $x = 1$.

   We now examine the intervals between the zeros.

Interval	$x < \dfrac{-4}{3}$	$\dfrac{-4}{3} < x < 1$	$x > 1$
$f'(x)$	$f'(-2) < 0$	$f'(0) > 0$	$f'(2) < 0$

   f is decreasing before $x = \dfrac{-4}{3}$, then increasing until $x = 1$, then decreasing again.

2. Local maximum at $x = -4$. Local minimum at $x = 3$.

   Since $h'(x)$ changes from positive to negative at $x = -4$, h goes from increasing to decreasing, creating a local maximum. Since $h'(x)$ changes from negative to positive at $x = 3$, h goes from decreasing to increasing, creating a local minimum. Note that at $x = -1$, $h'(x)$ does not change signs, so h has no extreme value there.

3. Absolute maximum of $e - 1$ at $x = \dfrac{1}{e}$. Absolute minimum of 1 at $x = 1$.

   Find the local extreme values and compare them to the endpoints.

   $g(x) = \ln(x) + \dfrac{1}{x}$, so $g'(x) = \dfrac{1}{x} - \dfrac{1}{x^2} = 0$ (or ∞) will give critical points.

   $\dfrac{x-1}{x^2} = 0 \Rightarrow x = 1$ is a critical point. $x = 0$ is not in the domain of the problem.

   $g''(x) = \dfrac{-1}{x^2} + \dfrac{2}{x^3}$

   $g''(1) = -1 + 2 = 1$. Since $g''(1) > 0$, g has a local minimum at $x = 1$.

   $g(1) = \ln(1) + \dfrac{1}{1} = 1$

   Now check the endpoints,

   $g\left(\dfrac{1}{e}\right) = \ln\left(\dfrac{1}{e}\right) + e = -1 + e \approx 1.7$, and $g(e) = \ln(e) + \dfrac{1}{e} \approx 1.4$.

   So g has an absolute maximum of $e - 1$ at $x = \dfrac{1}{e}$ and an absolute minimum of 1 at $x = 1$.

4. f is concave up on $(-\infty, -1)$ and concave down on $(-1, \infty)$.

   Find the intervals where $f''(x)$ is positive and negative.

   $f(x) = \tan^{-1}(x + 1)$, and $f'(x) = \dfrac{1}{1+(x+1)^2} = \dfrac{1}{x^2 + 2x + 2}$.

   $f'(x) = (x^2 + 2x + 2)^{-1}$, so $f''(x) = -1(x^2 + 2x + 2)^{-2}(2x + 2)$.

   $\dfrac{-(2x+2)}{(x^2+2x+2)^2} = 0 \Rightarrow x = -1$ is a possible point of inflection.

   The denominator of $f''(x)$ is always positive, so check the sign of the numerator.

$x < -1$	$x > -1$
$f''(-2) > 0$	$f''(0) < 0$
Concave up	Concave down

5. Since f is differentiable over the real numbers, it is also continuous. Since $f(2) = 4.5 = f(5)$, the average rate of change on the interval $[2, 5]$ is 0. The Mean Value Theorem guarantees at least one $x = c$ in the interval $(2, 5)$ where $f'(c) = 0$, giving f a horizontal tangent.

6. $c = \dfrac{64}{27}$

$g(8) = 8^{\left(\frac{2}{3}\right)} = 4$, and $g(0) = 0$.

$g(x) = x^{\left(\frac{2}{3}\right)}$, so $g'(x) = \dfrac{2}{3} \, x^{\left(\frac{-1}{3}\right)} = \dfrac{2}{3\sqrt[3]{x}}$.

$\dfrac{2}{3\sqrt[3]{c}} = \dfrac{g(8) - g(0)}{8 - 0} = \dfrac{1}{2}$

$\sqrt[3]{c} = \dfrac{4}{3} \Rightarrow c = \dfrac{64}{27}$ [c lies in the interval $(0, 8)$.]

7. h is increasing on $[-1, 0]$, $[0, 2)$, and $[8, \infty)$. h is decreasing on $(-\infty, -1]$ and $(2, 8]$.

$h'(x) = \dfrac{x^2(x-8)(x+1)}{x-2} = 0 \Rightarrow x = 0$ or $x = 8$ or $x = -1$ are critical points.

$x = 2$ is also a critical point because it makes $h'(x)$ undefined.

Examine $h'(x)$ in each interval.

$x < -1$	$-1 < x < 0$	$0 < x < 2$	$2 < x < 8$	$x > 8$
$h'(-2) < 0$	$h'\left(\frac{-1}{2}\right) > 0$	$h'(1) > 0$	$h'(5) < 0$	$h'(9) > 0$
h decreasing.	h increasing.	h increasing.	h decreasing.	h increasing.

(number line marked at 1, 0, 2, 8)

8. Inflection points at $x = 0$ and $x = 2$.

$f'(x) = x^2(x - 3) = x^3 - 3x^2$, so $f''(x) = 3x^2 - 6x$

$3x^2 - 6x = 0$

$3x(x - 2) = 0 \Rightarrow$ $x = 0$ or $x = 2$ are possible inflection points.

Reason graphically: $f''(x)$ is a parabola that opens upward and crosses the x-axis at 0 and 2. Since f'' changes signs, f has inflection points at 0 and 2.

9. h has a local maximum at about $x = 3$ and a local minimum at about $x = 5.5$. Since $h'(x)$ changes from positive to negative at $x = 3$, h goes from increasing to decreasing and has a local maximum there. Since $h'(x)$ changes from negative to positive at about $x = 5.5$, h goes from decreasing to increasing and has a local minimum there. Nothing can be determined about the endpoints, where $h'(x)$ is not defined.

10. h is concave up on $(0, 1)$ and $(4, 6)$. Referring to Figure 5.29, since h' is increasing on the intervals $(0, 1)$ and $(4, 6)$, its derivative h'' is positive. When h'' is positive, h is concave up.

11. $L(x) = -1(x - 5) + 3$

Reading off the graph of Figure 5.29, $h'(5) \approx -1$. This is the slope of h at $x = 5$.

Using $h(5) = 3$, the point-slope form of $h(x)$ is $y - 3 = -1(x - 5)$, or $L(x) = -1(x - 5) + 3$.

12.

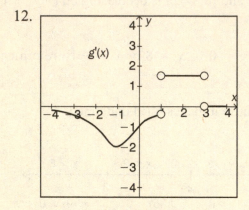

Figure 5.31

Figure 5.31 is the graph of $g'(x)$. or the slope of g. g (see Figure 5.30) starts with a small negative slope at $x = -4$, which remains negative until $x = 1$. The slope of g is most negative at the inflection point near $(-1, 0)$. From $x = 1$ to $x = 3$, g is linear with a constant slope of $\dfrac{3}{2}$. From $x = 3$ to $x = 4$, g is horizontal and has a slope of 0. There are open circles at $x = 1$ and $x = 3$ because g has corners so $g'(x)$ is undefined.

13. $t = \dfrac{3}{2}$ seconds and $t = \dfrac{9}{2}$ seconds

The particle is stationary when its velocity, $x'(t)$ equals 0.

$$x(t) = \sin\left(\frac{\pi}{3}t\right), \text{ so } x'(t) = v(t) = \frac{\pi}{3}\cos\left(\frac{\pi}{3}t\right).$$

$$x'(t) = \frac{\pi}{3}\cos\left(\frac{\pi}{3}t\right) = 0 \text{ when } \frac{\pi}{3}t = \frac{\pi}{2} + k\pi, \text{ where } k \text{ is any integer.}$$

Solve on the domain $0 \le t \le 5$.

$$\frac{\pi}{3}t = \frac{\pi}{2} \Rightarrow t = \frac{3}{2} \text{ or } \frac{\pi}{3}t = \frac{3\pi}{2} \Rightarrow t = \frac{9}{2}$$

14. Radius = 2 inches and height = 4 inches

Minimize surface area with a fixed volume of 16π.

$$S = 2\pi r^2 + 2\pi rh, \text{ and } V = \pi r^2 h \Rightarrow 16\pi \Rightarrow h = \frac{16}{r^2}$$

Substituting h, $S = 2\pi r^2 + 2\pi r \cdot \dfrac{16}{r^2} = 2\pi r^2 + 32\pi r^{-1}$.

$$\frac{dS}{dr} = 4\pi r - 32\pi r^{-2}$$

$$4\pi r - 32\pi r^{-2} = 0$$

$$4\pi r^{-2}(r^3 - 8) = 0 \Rightarrow r = 2, \text{ and } h = \frac{16}{(2)^2} = 4.$$

15. $\dfrac{64}{3}$ cubic inches

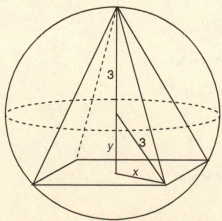

Figure 5.32

Sketch the problem, as shown in Figure 5.32.

$V_{\text{pyramid}} = \dfrac{1}{3} Bh$, where B = base area.

Use $A_{\text{square}} = \dfrac{1}{2} d^2$, where d = diagonal.

Using the right triangle shown,

$x = \sqrt{9 - y^2} \implies d = 2\sqrt{9 - y^2}$.

$V = \dfrac{1}{3}\left[\dfrac{1}{2} d^2\right] h$, so $V = \dfrac{1}{3}[2(9 - y^2)](y + 3)$.

$V = \dfrac{2}{3}(-y^3 - 3y^2 + 9y + 27)$, so $V' = \dfrac{2}{3}(-3y^2 - 6y + 9) = -2(y^2 + 2y - 3)$.

$V' = 0 \implies -2(y + 3)(y - 1) = 0$.

$y = 1$, and $V(1) = \dfrac{2}{3}(-1^3 - 3 \cdot 1^2 + 9 \cdot 1 + 27) = \dfrac{64}{3}$ cubic inches

16. $\dfrac{9}{4}$ feet per second

By the Pythagorean Theorem, with x = distance from the base of the ladder to the wall, and y = distance from the top of the ladder to the floor, $x^2 + y^2 = 10^2$.

When $y = 6$, $x = 8$.

Differentiate with respect to time, noticing the length of the ladder is constant, so its derivative equals 0. Also, since the ladder is sliding down the wall, $\dfrac{dy}{dt} = -3$.

$2x\dfrac{dx}{dt} + 2y\dfrac{dy}{dt} = 0$

Substitute known values.

$2(8)\dfrac{dx}{dt} + 2(6)(3) = 0$

$16\dfrac{dx}{dt} = 36$

$\dfrac{dx}{dt} = \dfrac{9}{4}$ feet per second

17. $4\sqrt{3}$ meters per second

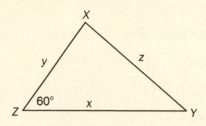

Figure 5.33

Sketch the problem, as in Figure 5.33

$\dfrac{dy}{dt} = 4$ meters per second, and $\dfrac{dx}{dt} = 8$ meters per second.
In 5 seconds,

$y = 20$ meters, and $x = 40$ meters.

The Law of Cosines relates the variables.

$z^2 = x^2 + y^2 - 2xy\cos(Z)$

$z^2 = 40^2 + 20^2 - 2(40)(20)\cos(60)$

$z^2 = 1600 + 400 - 1600\left(\dfrac{1}{2}\right)$

$z^2 = 1200 \implies z = 20\sqrt{3}$

Using implicit differentiation,

$2z\,\dfrac{dz}{dt} = 2x\,\dfrac{dx}{dt} + 2y\,\dfrac{dy}{dt} - 2\cos(Z)\left(x\,\dfrac{dy}{dt} + y\,\dfrac{dx}{dt}\right)$

Substituting known values,

$2(20\sqrt{3})\,\dfrac{dz}{dt} = 2(40)(8) + 2(20)(4) - 2\left(\dfrac{1}{2}\right)(40\cdot 4 + 20\cdot 8)$

$40\sqrt{3}\,\dfrac{dz}{dt} = 480$

$\dfrac{dz}{dt} = \dfrac{12}{\sqrt{3}} = 4\sqrt{3}$ meters per second

18. The difference in areas is 2.4π square inches.

$A = \pi r^2$, so $A'(r) = 2\pi r$.

$A'(6) = 12\pi$, and $A(6) = 36\pi$.

$L(r) = 36\pi + 12\pi(r - 6)$

$L(r) = 36\pi + 12\pi(6.2 - 6)$

$= 36\pi + 12\pi(0.2)$

$= 36\pi + 2.4\pi$

The estimated change in the area is the difference between the result and the original area of 36π. The difference is 2.4π.

As a point of interest and connection of ideas, the estimated difference could actually have been found quite easily by using differentials:

$dA = 2\pi r \cdot dr$, so $dA = 2\pi(6)(0.2) = 2.4\pi$.

19. $\sin\left(\dfrac{\pi}{6}\right) \approx 0.523$

$g(0) = \sin(0) = 0$, so the point of tangency is $(0, 0)$.

$g(x) = \sin\left(\dfrac{x}{2}\right)$, so $g'(x) = \dfrac{1}{2}\cos\left(\dfrac{x}{2}\right)$ and $g'(0) = \dfrac{1}{2}\cos(0) = \dfrac{1}{2}$.

$L(x) = 0 + \dfrac{1}{2}(x - 0) = \dfrac{1}{2}x$

Since $g(x) \approx L(x)$, $\sin\left(\dfrac{x}{2}\right) \approx \dfrac{1}{2}x$ for x-values near 0.

Estimating $\sin\left(\dfrac{\pi}{6}\right)$ requires substituting $x = \dfrac{\pi}{3}$ into the equivalence.

$\sin\left(\dfrac{\frac{\pi}{3}}{2}\right) \approx \dfrac{1}{2}\left(\dfrac{\pi}{3}\right) = \dfrac{\pi}{6}$, so $\sin\left(\dfrac{\pi}{6}\right) \approx 0.523$. Of course, the true value is 0.5.

20. $\dfrac{1}{2}$

$\displaystyle\lim_{x\to 1}\dfrac{\cos(\pi x)+x}{x^2-1}$ has the form $\dfrac{0}{0}$, so apply L'Hôpital's Rule.

$$\lim_{x\to 1}\dfrac{\cos(\pi x)+x}{x^2-1}=\lim_{x\to 1}\dfrac{-\pi\sin(\pi x)+1}{2x}$$

$$=\dfrac{-\pi\sin(\pi)+1}{2(1)}$$

$$=\dfrac{1}{2}$$

21. 0

$\displaystyle\lim_{x\to\infty}\dfrac{\ln(3x)}{\sqrt{x}}$ has the form $\dfrac{\infty}{\infty}$, so apply L'Hôpital's Rule.

$$\lim_{x\to\infty}\dfrac{\ln(3x)}{\sqrt{x}}=\lim_{x\to\infty}\left(\dfrac{\frac{3}{(3x)}}{\frac{1}{(2\sqrt{x})}}\right)$$

$$=\lim_{x\to\infty}\dfrac{1}{x}\cdot\dfrac{2\sqrt{x}}{1}$$

$$=\lim_{x\to\infty}\dfrac{2}{\sqrt{x}}$$

$$=0$$

CHAPTER 6

Antidifferentiation and Definite Integrals

CHAPTER 6

ANTIDIFFERENTIATION AND DEFINITE INTEGRALS

6.1　INTRODUCTION

To this point the entire focus of the material has been on taking derivatives of functions and utilizing this process to achieve a better understanding of the behavior of functions as well as finding solutions to applied problems. The rest of an introductory calculus course reverses this process by using antidifferentiation and explores the multitude of practical situations for which this is useful. The connection between differential and integral calculus is one of the most amazing and beautiful achievements in mathematics.

6.2　CONCEPT OF THE ANTIDERIVATIVE

Consider a simple function such as $f(x) = 2x$. This function could very easily be the derivative of another function, such that $\dfrac{dF}{dx} = 2x$. From previous experience, it should be relatively easy to conclude that $F(x)$ could be x^2. But could it also be $F(x) = x^2 + 3$? How about $F(x) = x^2 - 10$? Of course it could, since the derivative of a constant is 0.

Antiderivative

A function $F(x)$ is an antiderivative of a function $f(x)$ if $\dfrac{dF}{dx} = f(x)$.

All antiderivatives of a given function differ by a constant.

At this point, let's examine a few examples just to get the idea of finding a simple antiderivative. It is wise to recognize that not every antiderivative is simple to find, and some functions have no symbolic antiderivative at all. This will be examined in greater detail later.

EXAMPLE 6.1

Find an antiderivative of $y' = 4x^3$.

SOLUTION

At this point, use a guess and check method. Since y' is a polynomial, its derivative came from reducing its power by 1, so guess $y = x^4$. Sure enough, if $y = x^4$ or $y = x^4 +$ (any constant), $y' = \dfrac{dy}{dx} = 4x^3$.

EXAMPLE 6.2

Find a representation of all functions whose derivative is $f(x) = \sin(3x)$.

SOLUTION

A first guess may be $F(x) = \cos(3x)$ because the derivative of cosine involves sine. But remember that the derivative of cosine is the opposite of sine, and that there should be a chain rule factor from the derivative of the angle, $3x$. Thus, if $F(x) = \cos(3x)$, then $\dfrac{dF}{dx} = -\sin(3x)\dfrac{d}{dx}(3x) = -3\sin(3x)$.

Since the 3 and the negative sign are not present in the given function $f(x)$, something must have canceled them. Divide the original guess by -3 and remember the addition of a constant.

Now $F(x) = \dfrac{\cos(3x)}{-3} + C$.

Checking, $\dfrac{dF}{dx} = \dfrac{d}{dx}\left(\dfrac{\cos(3x)}{-3} + C\right)$

$= \dfrac{-3\sin(3x)}{-3} + 0$

$= \sin(3x)$

The symbol for finding an antiderivative is the integral symbol $\int$. Its origin will be explored shortly, but for the moment, $\int f(x)\,dx = F(x) + C$, where $F(x)$ is an antiderivative of f and C is any constant. $f(x)dx$ is called the integrand. An integral in this form is called an indefinite integral.

EXAMPLE 6.3

Find $\int e^{2x}\,dx$.

SOLUTION

$$\int e^{2x}\,dx = \frac{1}{2} e^{2x} + C$$

The $\frac{1}{2}$ is necessary to cancel the chain rule factor of 2 which would be produced by differentiating the exponent.

ANTIDIFFERENTIATION FORMULAS

The first few examples of this chapter determined antiderivatives by a trial and revise method, but there actually is some organization to the process. Most of the basic derivative formulas have a corresponding antiderivative formula that "reverses" the process to find an antiderivative. Some of these formulas are given in Table 6.1 and Table 6.2. Just like the differentiation formulas, they should be committed to memory. In each table, u is any arbitrary function, and C, n, and a are constants.

Table 6.1

	Differentiation Formula	Indefinite Integral Formula
1.	$d(u^n) = n \cdot u^{n-1}du$ (n is any real number)	$\int u^n\,du = \dfrac{u^{n+1}}{n+1} + C,\ n \neq -1$
2.	$d[\ln(u)] = \dfrac{1}{u}\,du$	$\int \dfrac{du}{u} = \ln|u| + C$
3.	$d(e^u) = e^u du$	$\int e^u\,du = e^u + C$
4.	$d(a^u) = a^u \cdot \ln(a)du$	$\int a^u\,du = \dfrac{a^u}{\ln(a)} + C$

(continued)

Table 6.1 (*continued*)

	Differentiation Formula	Indefinite Integral Formula
5.	$d[\sin(u)] = \cos(u)du$	$\int \cos(u)\,du = \sin(u) + C$
6.	$d[\cos(u)] = 2\sin(u)du$	$\int \sin(u)\,du = -\cos(u) + C$
7.	$d[\tan(u)] = \sec^2(u)du$	$\int \sec^2(u)\,du = \tan(u) + C$
8.	$d[\cot(u)] = -\csc^2(u)du$	$\int \csc^2(u)\,du = -\cot(u) + C$
9.	$d[\sec(u)] = \sec(u)\tan(u)du$	$\int \sec(u)\tan(u)\,du = \sec(u) + C$
10.	$d[\csc(u)] = -\csc(u)\cot(u)du$	$\int \csc(u)\cot(u)\,du = -\csc(u) + C$

Lest we forget the inverse trigonometric functions, Table 6.2 completes the list! Interestingly, since the derivatives of the inverse cofunctions such as arctan and arccot differ only in sign, there is no need for a special antiderivative formula for each.

Table 6.2

	Differentiation Formula	Indefinite Integral Formula
11.	$d[\sin^{-1}(u)] = \dfrac{du}{\sqrt{1-u^2}}$	$\int \dfrac{du}{\sqrt{1-u^2}} = \sin^{-1}(u) + C$
12.	$d[\tan^{-1}(u)] = \dfrac{du}{1+u^2}$	$\int \dfrac{du}{1+u^2} = \tan^{-1}(u) + C$
13.	$d[\sec^{-1}(u)] = \dfrac{du}{|u|\sqrt{u^2-1}}$	$\int \dfrac{du}{|u|\sqrt{u^2-1}} = \sec^{-1}(u) + C$

EXAMPLE 6.4

$$\int 5u^3\,du$$

SOLUTION

$$\int 5u^3\,du = 5\frac{u^{3+1}}{3+1} + C$$
$$= \frac{5}{4}u^4 + C$$

EXAMPLE 6.5

$$\int \frac{dy}{\sqrt{1-y^2}}, \, |y| < 1$$

SOLUTION

$$\int \frac{dy}{\sqrt{1-y^2}} = \sin^{-1}(y) + c$$

ANTIDIFFERENTIATION BY *U*-SUBSTITUTION

Of course, not all antiderivatives are going to be set up so nicely to fit the formulas. This is where the mathematician's skill and previous knowledge come into play. The more ingrained the derivative formulas are, the easier antidifferentiation will be. The method of *u*-substitution (named for no other reason than the common use of the variable *u*) is a critical tool for organizing the process of antidifferentiation, and seeing more clearly which formula fits which situation. To provide one more tool for working with indefinite integrals, some intuitive properties of indefinite integrals are presented below.

Integral of a Sum Property

The integral of a sum is the sum of the integrals.

$$\int [u(x) + v(x)]\, dx = \int u(x)\, dx + \int v(x)\, dx$$

Constant Multiple Property

The integral of a constant multiple of a function is the constant times the integral.

Let k be a constant. $\int k \cdot u(x)\, dx = k \int u(x)\, dx$

These two properties are very useful in rewriting integrals in forms that will make it easier to make an association with the indefinite integral formulas.

The key idea in *u*-substitution is to look, within the integrand, for a function and any constant multiple of that function's derivative. The function will be equated with *u* and its derivative will be equated with *du*. After substituting for each part of the integrand, the form of the integral will be a constant multiple of one of the given antidifferentiation formulas.

EXAMPLE 6.6

Use *u*-substitution to rewrite and antidifferentiate $\int x^2(x^3+1)^5\, dx$.

SOLUTION

Let $u = x^3 + 1$ because its derivative is a constant multiple of x^2 which is also part of the integrand.

Differentiating both sides of the equation, $u = x^3 + 1 \Rightarrow du = 3x^2 dx$.

There are a variety of methods to substitute. Two of those will be presented side by side here for comparison.

Isolate the exact terms that appear as factors in the integrand.	Match the terms of the integrand to *u* and *du* by multiplication and balancing. The integrand needs a 3 to match *du*.

$x^3 + 1 = u$ and

$$du = 3x^2 dx \Rightarrow \frac{1}{3}du = x^2 dx$$

Substitute to get

$$\int x^2(x^3+1)^5\, dx = \int u^5 \cdot \frac{1}{3}\, du$$

$$= \frac{1}{3}\int u^5\, du$$

Multiply by 3 and by $\frac{1}{3}$, then substitute.

$$\int x^2(x^3+1)^5\, dx = \frac{1}{3}\int 3x^2(x^3+1)^5\, dx$$

$$= \frac{1}{3}\int u^5\, du$$

Each now uses the formula $\int u^n\, du = \dfrac{u^{n+1}}{n+1}+C$ to get

$$\frac{1}{3}\int u^5\, du = \frac{1}{3}\cdot\frac{u^6}{6}+C$$

$$= \frac{1}{18}(x^3+1)^6 + C$$

It is highly recommended that you choose the method with which you are most comfortable and master that method. Either one achieves the same goal.

EXAMPLE 6.7

Use u-substitution to rewrite and antidifferentiate $\int \dfrac{2x+1}{5x^2+5x}\,dx$.

SOLUTION

Let $u = 5x^2 + 5x$. This choice is made because the numerator contains a linear function that is the derivative of a constant function.

$u = 5x^2 + 5x \Rightarrow du = (10x + 5)dx$. Notice $10x + 5$ is a multiple of $2x + 1$.

So $du = (10x + 5)dx$ transforms to $\dfrac{1}{5}du = (2x+1)dx$.

Substituting into the integrand,

$$\int \frac{2x+1}{5x^2+5x}\,dx = \int \frac{\left(\frac{1}{5}\right)du}{u}$$

$$= \frac{1}{5}\int \frac{du}{u}$$

$$= \frac{1}{5}\ln|u| + C$$

$$= \frac{1}{5}\ln|5x^2 + 5x| + C$$

EXAMPLE 6.8

Use u-substitution to rewrite and antidifferentiate $\int 8e^{\sin(2x)}\cos(2x)\,dx$.

SOLUTION

Let $u = \sin(2x)$, so $du = 2\cos(2x)dx \Rightarrow 4du = 8\cos(2x)dx$

$$\int 8e^{\sin(2x)}\cos(2x)\,dx = \int 4e^u\,du$$

$$= 4\int e^u\,du$$

$$= 4e^u + C$$

$$= 4e^{\sin(2x)} + C$$

ANALYTIC REWRITING OF INTEGRANDS

Not all integrals fit nicely into the *u*-substitution method. Sometimes analytic changes must be made to the integrand to make it easier to find the antiderivative. Those analytic changes can include, but not be limited to, expanding a product, dividing out a rational function, or even using a trigonometric identity. As a general rule, *u*-substitution should be the first method considered, but if that does not work, try rewriting the integrand in a different form. In addition, be sure that no substitution results in the differential, *du*, being in the denominator of a fraction or being raised to any power.

EXAMPLE 6.9

$$\int \frac{x^2+2}{x^2+1}\,dx$$

SOLUTION

$$\frac{x^2+2}{x^2+1} = \frac{x^2+1+1}{x^2+1}$$

$$= \frac{x^2+1}{x^2+1} + \frac{1}{x^2+1}$$

$$= 1 + \frac{1}{x^2+1}$$

$$\int \frac{x^2+2}{x^2+1}\,dx = \int \left(1 + \frac{1}{x^2+1}\right) dx$$

$$= \int 1\,dx + \int \frac{1}{x^2+1}\,dx$$

$$= x + \tan^{-1}(x) + C$$

EXAMPLE 6.10

$$\int \frac{\cos(4x)}{\cot(4x)}\,dx$$

SOLUTION

$$\frac{\cos(4x)}{\cot(4x)} = \cos(4x)\tan(4x)$$

$$= \cos(4x)\frac{\sin(4x)}{\cos(4x)} = \sin(4x)$$

$$\int \frac{\cos(4x)}{\cot(4x)}\,dx = \int \sin(4x)\,dx$$

$$= -\frac{1}{4}\cos(4x) + C$$

In general, antidifferentiation tends to be a bit more difficult than differentiation for a few reasons. The methods are slightly more varied, there is a bit more trial and revision required, and u-substitution requires the use of differentiation to generate du, which then leads to antidifferentiation. Do not be discouraged if mastery comes slowly. Practice, practice, and more practice is the key to success. Just forge ahead making an initial conjecture about u and du, then revise your choice if necessary. For now, let's diverge momentarily from antidifferentiation to lay the groundwork to develop a second category of integral, the definite integral. Its foundation lies in numerical approximation and limits.

6.3 NUMERICAL APPROXIMATION

AREA UNDER A CURVE

If a cyclist rides a bike at a constant rate of 12 miles per hour for 2 hours, the distance traveled is 24 miles. Taking a look at a velocity versus time graph sheds a new light on this very simple problem and opens the door to some fascinating mathematics. Of course, "rate times time" was used to determine distance, but notice that in Figure 6.1 the rate is the vertical component of the graph and the time is the horizontal component. So multiplying rate times time is equivalent to multiplying height times width, which is the area of the rectangle for which the velocity graph is the upper edge. The area under the graph of velocity produced distance traveled. A very common and important concept in calculus is that the area under the graph of a rate versus time graph represents an accumulation of the quantity the rate is measuring. In Figure 6.1, that quantity is miles.

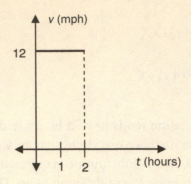

Figure 6.1

Naturally, whatever rate is being measured does not need to be constant. Consider slowly opening a water faucet such that the rate of flow of the water increases linearly. Suppose over 3 seconds the rate went from 0 ounces per second to 4 ounces per second. The graph of flow rate versus time would look different from the previous scenario, but the work to find the area under the graph would still be relatively simple. Figure 6.2 is a graph of the situation. If r is the rate of water flow in ounces per second (ops) and t is time in seconds, the area under the linear graph is simply

$$\frac{1}{2} \cdot 3 \text{ secs} \cdot 4 \frac{\text{ounces}}{\text{second}} = 6 \text{ ounces}$$

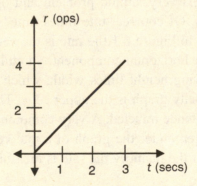

Figure 6.2

But what if the curve under which area is desired is neither constant nor linear? How would one calculate the area under a parabola? For the moment, let's be content with estimating that area using a method called the Rectangular Approximation Method.

RECTANGULAR APPROXIMATION METHOD

If a positive function is nonlinear and the area between that function and the x-axis is desired, it is approximated by using a designated number of rectangles drawn beneath the graph of the curve from the function to the x-axis. This is called the Rectangular Approximation Method, or RAM for short. Most students have actually done this previously by simply counting squares under a curve drawn on square-grid graph paper. In essence, that is the process for RAM, but greater sophistication can be introduced by using function values and limits, which lead to a much improved result!

On a closed interval [a, b], the number of equally wide rectangles will determine the width of each rectangle. The height of each rectangle can be determined by the function value of any x in each interval, but traditionally the left endpoint, right endpoint or midpoint value of a given interval is used. This text will use LRAM as the abbreviation for left-oriented rectangles, RRAM for right-oriented rectangles, and MRAM for midpoint-oriented rectangles.

EXAMPLE 6.11

Use four rectangles whose height is based on the left endpoint of each interval to estimate the area under $f(x) = x^2 + 3$ on the interval [0, 4].

SOLUTION

Dividing [0, 4] into 4 equally wide rectangles makes each rectangle 1 unit wide, as seen in Figure 6.3. The height of each rectangle here is based on the left endpoints, $f(0), f(1), f(2),$ and $f(3)$.

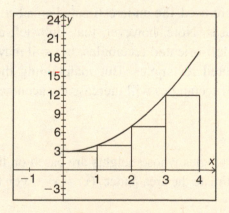

Figure 6.3

$$\text{Area} \approx f(0) \cdot 1 + f(1) \cdot 1 + f(2) \cdot 1 + f(3) \cdot 1$$
$$= 3 + 4 + 7 + 12$$
$$= 26$$

The answer to Example 6.11 has no units since the example is unrelated to any application. Additionally, recognize that it is merely an estimate. One can easily see the underestimated areas as the spaces above the rectangles and below the curve.

EXAMPLE 6.12

Calculate MRAM using four equal subdivisions on $f(x) = x^2 + 3$ on the interval $[0, 4]$.

SOLUTION

The curve is the same as that shown in Figure 6.3. The width of each rectangle is still 1 unit, but the heights now are based on the midpoint of each of the intervals $[0, 1]$, $[1, 2]$, $[2, 3]$, and $[3, 4]$.

$$\text{Area} \approx f(0.5) \cdot 1 + f(1.5) \cdot 1 + f(2.5) \cdot 1 + f(3.5) \cdot 1$$
$$= 3.25 + 5.25 + 9.25 + 15.25$$
$$= 33$$

You may suspect that this approximation is somewhat better than the left-oriented underestimate, and you are correct. But it can still be improved. So what could increase the accuracy in the previous examples? More rectangles! If eight rectangles are used, the underestimated area between the rectangles and the function decreases. Note, however, that eight left-oriented rectangles will be better than four left-oriented rectangles, but still may not be better than just four midpoint-oriented rectangles. But maintaining the same orientation, an increased number of rectangles will increase the accuracy of an approximation.

EXAMPLE 6.13

Use eight rectangles whose heights are based on the left endpoint of each interval to estimate the area under $f(x) = x^2 + 3$ on the interval $[0, 4]$.

SOLUTION

This is the same curve as in the two previous examples. Dividing [0, 4] into eight equally wide rectangles makes each rectangle $\frac{1}{2}$ unit wide (see Figure 6.4). The height of each rectangle is based on the function value at the left end of each of the eight intervals, $f(0), f(.5), f(1), f(1.5), ..., f(3.5)$. Notice that $f(4)$ is not used.

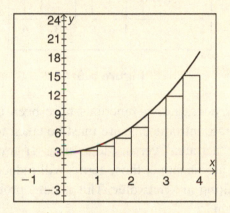

Figure 6.4

Area $\approx f(0)\cdot(.5) + f(.5)\cdot(.5) + f(1)\cdot(.5) + f(1.5)\cdot(.5) + ... + f(3.5)\cdot(.5)$

$= [3 + 3.25 + 4 + 5.25 + ... + 15.25](.5)$

$= 29.5$

Again, this is an underestimate since there is space above the rectangles and below the curve that is not included.

EXAMPLE 6.14

Calculate RRAM for $f(x) = x^2 + 3$ using eight rectangles.

SOLUTION

In a similar fashion, our calculation is

Area $\approx f(.5)(.5) + f(1)(.5) + ... + f(3.5)(.5) + f(4)(.5) = 37.5$

Example 6.13 underestimated the area at 29.5. This example overestimates the area at 37.5. It will eventually be shown that the actual area is $33\frac{1}{3}$.

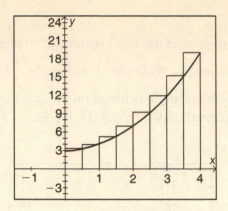

Figure 6.5

To this point only nonnegative functions have been used. When functions with negative values are introduced, care must be taken to distinguish between what one could call "net area" versus "total area." If any of the values of $f(x)$ in the Riemann sum are negative, those terms will reduce the accumulation of rectangular area resulting in a net value. This is not a problem as long as that is the result that is sought.

Example 6.15 shows how these two different results could each have significance. The mechanics of the Riemann sum are assumed to have been done in this example, so the focus is on the concept of the different results.

EXAMPLE 6.15

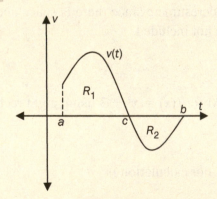

Figure 6.6

Let the graph of $v(t)$ in Figure 6.6 represent the velocity of a particle moving along the x-axis. When velocity is positive, the particle is moving right. Negative velocity indicates movement to the left. Suppose Riemann sums had been calculated on the graph of $v(t)$ from a to c, and from c to b. If the result of each sum had been $R_1 = 12$ and $R_2 = -7$, find the total distance traveled and the displacement of the particle during the time interval $[a, b]$.

SOLUTION

The negative result in R_2 would have come from multiplying negative velocity values times small positive changes in time. $R_2 = -7$ means the particle moved left 7 units during the interval $[c, b]$.

Total distance does not take into account direction, so the sign of R_2 must be ignored. Total distance $= R_2 + |R_2| = 19$.

Displacement, however, is the *net* change in position. Displacement $= R_1 + R_2 = 12 + (-7) = 5$. The particle moved 12 units right, 7 units left, and ended up 5 units right of its position at time a. If a Riemann sum had been done on the entire interval $[a, b]$, the result would have been displacement.

Example 6.15 shows that clarity is required when talking about area. The accepted standard is that area is always considered a positive quantity, but in context has significance when reported as a signed quantity. Additionally, it can certainly make sense to talk about "net area" since it has meaning, such as in the case here of displacement in linear motion.

TRAPEZOIDAL RULE

One other way to estimate area is to use trapezoids instead of rectangles. This is accomplished by connecting, for each subinterval, the left and right endpoints on the graph of the function, then calculating and summing a series of trapezoidal regions. Figure 6.7 shows one subinterval along a function to make the orientation of the trapezoid visually clear. Recall that the area of a trapezoid is $A = \dfrac{1}{2} h(b_1 + b_2)$, where b_1 and b_2 are the parallel sides, and h is the distance between them. In Figure 6.7, $h = x_1 - x_0 = \Delta x$. Δx is found the same way as with rectangular approximations. $b_1 = f(x_0)$ and $b_2 = f(x_1)$. So for the trapezoid in Figure 6.7, Area $= \dfrac{1}{2} h \cdot [f(x_0) + f(x_1)]$.

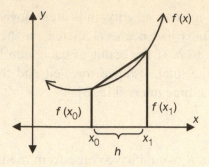

Figure 6.7

When multiple subintervals are used, the right base of one trapezoid also serves as the left base of the next trapezoid, so all bases on the interior of the domain actually get used twice.

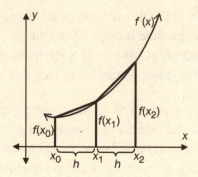

Figure 6.8

So the area in Figure 6.8 is

$$\text{Area} = \frac{1}{2}h \cdot [f(x_0) + f(x_1)] + \frac{1}{2}h \cdot [f(x_1) + f(x_2)]$$

Notice that $f(x_1)$ is used twice. Factoring out the common $\frac{1}{2}h$ gives

$$\text{Area} = \frac{1}{2}h \cdot [f(x_0) + f(x_1) + f(x_1) + f(x_2)]$$

$$= \frac{1}{2}h \cdot [f(x_0) + 2f(x_1) + f(x_2)]$$

Trapezoidal Rule

Let $f(x)$ be a nonnegative continuous function on the interval $[a, b]$. The area between $f(x)$ and the x-axis can be estimated with trapezoids using n equal subdivisions by

$$\text{Area} \approx \frac{h}{2} \cdot [f(x_0) + 2f(x_1) + 2f(x_2) + \ldots + 2f(x_{n-1}) + f(x_n)]$$

where

$$h = \Delta x = \frac{b-a}{n}.$$

If $f(x)$ takes on both positive and negative values in the interval, the trapezoidal rule will produce what has been discussed as net area.

Example 6.16 revisits $f(x) = x^2 + 3$ so that the result may be compared to prior rectangular approximations. Notice in Figure 6.9 that with just eight trapezoids, the slanted sides of the trapezoids connecting points on the function already become very difficult to see. This is actually a good thing because the spaces between the graph of the function and the slanted sides of the trapezoids that connect points on the function are what creates the error in the approximation, and now these spaces are very small.

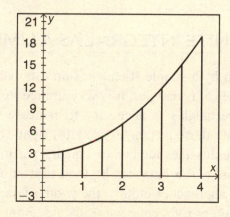

Figure 6.9

EXAMPLE 6.16

Use trapezoids to estimate the area under $f(x) = x^2 + 3$ and above the $0 = x$-axis on the interval $[0, 4]$ using eight equal subdivisions.

SOLUTION

Again, this is the same curve as in the previous examples.

$$h = \frac{4-0}{8} = \frac{1}{2} \text{ so } \frac{h}{2} = \frac{1}{4}$$

$$\text{Area} \approx \frac{1}{4}[f(0) + 2f(.5) + 2f(1) + \ldots + 2f(3.5) + f(4)]$$

$$= \frac{1}{4}[3 + 2(3.25) + 2(4) + \ldots + 2(15.25) + 19]$$

$$= 33.5$$

It is interesting to note that 33.5 is the average of the results acquired by using eight left-oriented rectangles (29.5; see Example 6.13) and eight right-oriented rectangles (37.5; see Example 6.14). This is no coincidence. In fact, using equal numbers of subdivisions, the trapezoidal approximation will always be the average of LRAM and RRAM. It will not, however, be related in any particular way with a midpoint approximation, MRAM.

6.4 THE DEFINITE INTEGRAL AS A LIMIT

An area calculation by a simple Riemann sum estimates a net area. With a relatively small number of rectangles, it is very limited in its accuracy. Increasing the number of rectangles is necessary to increase the accuracy of the approximation, whether those rectangles are left-, right-, or midpoint-oriented. Fortunately, with calculus the number of equally partitioned rectangles can increase without bound and a limit can be used to examine the result. As the number of rectangles approaches infinity, the width of each rectangle becomes infinitely small. The left or right endpoint or the midpoint is no longer required to determine the height of the rectangle. Any value of x in each subinterval may be used to determine the height. This is because the difference between the left, right, and midpoint function values approaches 0 as Δx approaches 0. The result of this limit is defined to be what is called the definite integral of a function on a given interval $[a, b]$.

Definite Integral

Let $f(x)$ be continuous on a given interval $[a, b]$. Let n be the number of rectangles used, so the width of each rectangle is $\Delta x = \dfrac{b-a}{n}$. $\lim\limits_{n \to \infty} \sum\limits_{i=1}^{n} f(x_i)\Delta x$ defines the definite integral of $f(x)$ on $[a, b]$. In symbols, in any interval $[a, b]$,

$$\lim\limits_{n \to \infty} \sum\limits_{i=1}^{n} f(x_i)\Delta x = \int_a^b f(x)dx.$$

Of course, there is similarity to indefinite integrals: $f(x)$ is the integrand and dx is the variable of integration. The main difference is that a definite integral has lower and upper limits, a and b, respectively. Another difference is that a definite integral is an accumulator of area, and that area has meaning as it relates to a particular application. As the introductory example to Section 6.3 showed, the area under the velocity graph was an accumulation of distance. It will help to keep in mind that a definite integral is really a sum of an infinite number of products, $f(x_i) \cdot dx$, and the signs of those products can be positive or negative.

Definite Integral as Area

If $f(x) \geq 0$ on $[a, b]$, then $\int_a^b f(x)dx$ is the total area under $f(x)$ and above the x-axis.

If $f(x)$ takes on both positive and negative values in $[a, b]$, then $\int_a^b f(x)dx$ is the net area between $f(x)$ and the x-axis.

The next few examples take a geometric approach to determining the value of a definite integral. Remember that this is a perfectly valid approach and is sometimes much more efficient than other methods, which will be discussed later.

EXAMPLE 6.17

Find the value of $\int_{-4}^{4} \sqrt{16-x^2} \, dx$.

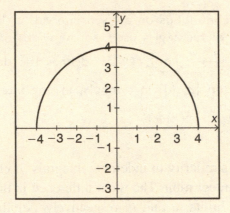

Figure 6.10

SOLUTION

The integral asks for the area under the graph of $f(x) = \sqrt{16-x^2}$ and the x-axis. The $f(x)$ is the equation of a semi-circle of radius 4 (see Figure 6.10). Of course, $f(x) \geq 0$, so $\int_{-4}^{4} \sqrt{16-x^2} \, dx = \frac{1}{2} \pi (4^2) = 8\pi$.

EXAMPLE 6.18

Evaluate $\int_{0}^{5} (4-2x) \, dx$.

SOLUTION

The function is shown in Figure 6.11. On [0, 2], $R_1 = \frac{1}{2}(4)(2) = 4$. R_2 lies below the x-axis and will count as "negative area." On [2, 5], $R_2 = \frac{1}{2}(3)(-6) = -9$. $\int_{0}^{5} (4-2x) \, dx = R_1 + R_2 = 4 + -9 = -5$.

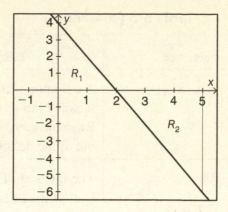

Figure 6.11

6.5   PROPERTIES OF DEFINITE INTEGRALS

Since definite integrals are defined as the limit of a Riemann sum, many of the properties of summation can be applied to definite integrals. Each property below is accompanied by a simple explanation of either why it works or what it means. $h(x)$ and $g(x)$ are continuous functions, and a, b, c, and k are real number constants.

Table 6.3

	Property	Explanation
1.	$\displaystyle\int_a^b h(x)\,dx = -\int_b^a h(x)\,dx$	Think of reversing the limits as summing from right to left, for which in a Riemann sum, Δx would be negative. Therefore, each product $h(x) \cdot \Delta x$ would be opposite the original product.
2.	$\displaystyle\int_a^a h(x)\,dx = 0$	There is no area under a function on an interval with no width.

(continued)

Table 6.3 (*continued*)

	Property	Explanation
3.	$\displaystyle\int_a^b k \cdot h(x)\,dx = k\int_a^b h(x)\,dx$	A common factor of k can be factored out of each product in a Riemann sum, so it can be factored out of the definite integral.
4.	$\displaystyle\int_a^b [h(x)+g(x)]\,dx =$ $\displaystyle\int_a^b h(x)\,dx + \int_a^b g(x)\,dx$	The integral of a sum equals the sum of the integrals.
5.	$\displaystyle\int_a^b h(x)\,dx + \int_b^c h(x)\,dx =$ $\displaystyle\int_a^c h(x)\,dx$	Areas of adjacent regions can be added.

EXAMPLE 6.19

Let $g(x)$ be continuous over all real numbers. If $\displaystyle\int_a^b g(x)\,dx = 11$ and $\displaystyle\int_c^a g(x)\,dx = 4$, find $\displaystyle\int_b^c g(x)\,dx$.

SOLUTION

If $\displaystyle\int_c^a g(x)\,dx = 4$, then $\displaystyle\int_a^c g(x)\,dx = -4$.

$$\int_a^b g(x)\,dx + \int_b^c g(x)\,dx = \int_a^c g(x)\,dx$$

$$11 + \int_b^c g(x)\,dx = -4 \Rightarrow \int_b^c g(x)\,dx = -15$$

EXAMPLE 6.20

If $\displaystyle\int_2^5 f(x)\,dx = 4$, evaluate $\displaystyle\int_2^5 [6f(x)-1]\,dx$.

SOLUTION

$$\int_2^5 [6f(x)-1]\,dx = \int_2^5 6f(x)\,dx - \int_2^5 1\,dx$$

Think of $\int_{2}^{5} 1\,dx$ as the area under $y = 1$ from $x = 2$ to $x = 5$, a 3 by 1 rectangle!

$$\int_{2}^{5} 6f(x)\,dx - \int_{2}^{5} 1\,dx = 6\int_{2}^{5} f(x)\,dx - \int_{2}^{5} 1\,dx$$
$$= 6(4) - 3(1)$$
$$= 21$$

EXAMPLE 6.21

If $b > a$ and $\int_{a}^{b} x^2\,dx = \frac{1}{3}b^3 - \frac{1}{3}a^3$, find an expression in terms of a and b for $\int_{a}^{b} (3x^2 + 2x - 1)\,dx$.

SOLUTION

This problem uses properties of definite integrals and a geometric approach to area.

$$\int_{a}^{b} (3x^2 + 2x - 1)\,dx = \int_{a}^{b} 3x^2\,dx + \int_{a}^{b} (2x - 1)\,dx$$
$$\int_{a}^{b} 3x^2\,dx = 3\int_{a}^{b} x^2\,dx$$
$$= 3\left[\frac{1}{3}b^3 - \frac{1}{3}a^3\right]$$
$$= b^3 - a^3$$

$\int_{a}^{b} (2x - 1)\,dx$ is shown in Figure 6.12.

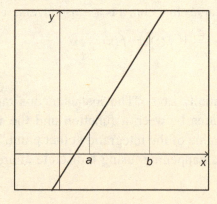

Figure 6.12

The area can be calculated as a trapezoid. The result does not depend on whether a and b are positive or negative.

The left base is $2a - 1$ and the right base is $2b - 1$. The height is $b - a$.

$$\begin{aligned} \text{Area} &= \frac{1}{2}(b-a)[2a-1+2b-1] \\ &= (b-a)[a+b-1] \\ &= b^2 - a^2 - b + a \end{aligned}$$

The sum of the two results is $b^3 - a^3 + b^2 - a^2 - b + a$.

6.6   FUNDAMENTAL THEOREM OF CALCULUS

One of the most impressive developments in calculus was the connection between the differential and integral branches found in what came to be called the Fundamental Theorem of Calculus. One part of the Fundamental Theorem asserts that a definite integral of a continuous function is a function of its upper limit and is therefore itself differentiable. A second part of the Fundamental Theorem establishes how to analytically evaluate a definite integral. Both parts will be presented here without formal proof, but if it enhances your understanding, any online search for the proof of the Fundamental Theorem of Calculus will produce an abundance of sources to read and study.

Fundamental Theorem of Calculus (Part 1)

If f is a continuous function, a is a constant, and $y = \int_a^x f(t)\,dt$,

then $\dfrac{d}{dx}(y) = \dfrac{d}{dx}\int_a^x f(t)\,dt = f(x)$.

This theorem essentially says, "The instantaneous rate of change of accumulation, or loss, of area between a function and the x-axis at any point, is equal to the function value of the integrand at that point." Let's take an intuitive way to explain what is happening, using a discrete approach to the concept of accumulating area under a curve. Think of a definite integral as a Riemann sum over a given interval. As the upper limit of the given interval changes, accumulated area increases or decreases by "adding" another infinitely thin rectangle to the previously summed rectangles. If the function values are "large" at a particular x value, then the next rectangle added will add more area than if the function values are small. It follows logically that the amount of change in accumulated area is related to the magnitude of the function being integrated.

Notice that the upper limit is simply x. How would this part of the theorem change if the upper limit was some function of x, say $h(x)$?

> If f is a continuous function, a is a constant, and $h(x)$ is a func-
> tion of x, then $\dfrac{d}{dx} \displaystyle\int_a^{h(x)} f(t)\, dt = f(h(x)) \cdot h'(x)$.

The Chain Rule helps to justify this result.

If $y = \displaystyle\int_a^{h(x)} f(t)\, dt$, write y as a composition of $y = \displaystyle\int_a^u f(t)\, dt$ and $u = h(x)$.

By the Fundamental Theorem, since y is a function of u, $\dfrac{dy}{du} = f(u)$.

Also $\dfrac{du}{dx} = h'(x)$. By the Chain Rule, $\dfrac{dy}{dx} = \dfrac{dy}{du} \cdot \dfrac{du}{dx}$.

Substituting gives $\dfrac{dy}{dx} = f(u) \cdot h'(x)$ and $u = h(x)$. Further substitution leads to the desired result, $\dfrac{dy}{dx} = f(h(x)) \cdot h'(x)$ or $\dfrac{d}{dx} \displaystyle\int_a^{h(x)} f(t)\, dt = f(h(x)) \cdot h'(x)$.

With a bit of practice the actual mechanics of using this part of the Fundamental Theorem is pretty elementary. It is important to understand that finding an antiderivative is not necessary prior to differentiating the definite integral function. It is also important to understand that the lower limit can be any constant and it will not change the result.

EXAMPLE 6.22

$$\frac{d}{dx}\int_3^x \sin(e^t)\,dt$$

SOLUTION

$$\frac{d}{dx}\int_3^x \sin(e^t)\,dt = \sin(e^x)$$

The Fundamental Theorem Part 1 was applied directly. An antiderivative function was not sought, and would be nontrivial to find.

EXAMPLE 6.23

If $y = \int_4^{x^2} e^{\sqrt{t}}\,\ln(t+1)dt$, find $\frac{dy}{dx}$.

SOLUTION

Let $u = x^2$ and $y = \int_4^u e^{\sqrt{t}}\,\ln(t+1)dt$.

$$\frac{du}{dx} = 2x \text{ and } \frac{dy}{du} = e^{\sqrt{u}}\,\ln(u+1)$$

$$\frac{dy}{dx} = \frac{dy}{du}\cdot\frac{du}{dx}$$

$$\frac{dy}{du} = e^{\sqrt{u}}\,\ln(u+1)\cdot 2x$$

$$= e^{\sqrt{x^2}}\,\ln(x^2+1)\cdot 2x$$

$$= e^{|x|}\,\ln(x^2+1)\cdot 2x$$

Note: A simpler way to get the result is to replace every t inside the integral with x^2, including the t in dt, which can be thought of as $d(x^2)$ or $2x$.

EXAMPLE 6.24

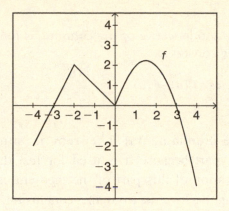

Figure 6.13

The graph of f on $[-4, 4]$ is shown in Figure 6.13. Let $g(x) = \int_{-2}^{x} f(t)\,dt$.

a) On what interval is $g(x)$ increasing?

b) Where does $g(x)$ have a local minimum?

c) On what interval(s) is $g(x)$ concave up?

SOLUTION

The answer to all three questions rests with the Fundamental Theorem. If $g(x) = \int_{-2}^{x} f(t)\,dt$, then $g'(x) = f(x)$, and $g''(x) = f'(x)$. The given graph of f is also a graph of $g'(x)$. This also means, the slope of f is $g'(x)$.

a) $g(x)$ is increasing on $[-3, 3]$ because $g'(x) = f(x) \geq 0$.

b) $g(x)$ has a local minimum at $x = -3$, because $g'(x) = f(x)$ changes from negative to zero to positive at $x = -3$, so g goes from decreasing to increasing. $g(x)$ also has a local minimum at $x = 4$, because $f(x) < 0$ on $(3, 4]$ meaning g decreases to the endpoint at $x = 4$.

c) $g(x)$ is concave up where $g''(x) = f'(x) > 0$. $f'(x) > 0$ when f is increasing. $g(x)$ is concave up on the intervals $(-4, -2)$ and $(0, 1.5)$.

Fundamental Theorem of Calculus (Part 2)

> Let F be any antiderivative of the continuous function f, and let a and b be constants.
>
> Then $\int_a^b f(x)\,dx = F(b) - F(a)$.

This part of the Fundamental Theorem is sometimes called the "Integral Evaluation" part because it is used for just that—to evaluate definite integrals. Justification of this part of the theorem can be achieved fairly succinctly by working backwards from the conclusion. It has already been established that a definite integral is a function of its upper limit, so let $H(x) = \int_a^x f(t)\,dt$. Also, if $F(x)$ is any other antiderivative of f, then it will differ from $H(x)$ by some arbitrary constant C, therefore $F(x) = H(x) + C$.

Begin with $F(b) - F(a)$

$$F(b) - F(a) = [H(b) + C] - [H(a) + C]$$
$$= \int_a^b f(t)\,dt - \int_a^a f(t)\,dt$$
$$= \int_a^b f(t)\,dt - 0$$
$$= \int_a^b f(t)\,dt$$

This means the process for evaluating a definite integral is now clearly defined. Find the antiderivative of the integrand, and then find the difference of the antiderivative evaluated at the upper limit minus the antiderivative evaluated at the lower limit. The result will be the net area (not total area) between the graph of the integrand and the x-axis.

EXAMPLE 6.25

Evaluate $\int_1^5 \dfrac{3}{x+1}\,dx$.

SOLUTION

$$\int_1^5 \frac{3}{x+1}\,dx = 3\int_1^5 \frac{dx}{x+1}$$
$$= 3\ln(x+1)\Big|_1^5$$
$$= 3[\ln(6)-\ln(2)]$$
$$= 3\ln(3)$$

EXAMPLE 6.26

Evaluate Example 6.11, examined in the section on the Rectangular Approximation Method, as $\int_0^4 (x^2+3)\,dx$.

SOLUTION

$$\int_0^4 (x^2+3)\,dx = \frac{x^3}{3}+3x\,\Big|_0^4$$
$$= \frac{4^3}{3}+3(4)-\left[\frac{0^3}{3}+3(0)\right]$$
$$= 33\frac{1}{3}$$

U-SUBSTITUTION REVISITED

When definite integrals become too complicated to find the antiderivative by guess and check, the *u*-substitution method can be used to evaluate them in a manner similar to that for indefinite integrals. There are two choices for dealing with the limits. The first, and less efficient, of the methods is to resubstitute so the antiderivative is in terms of the original variable, then evaluate using the given limits. The second choice is to change the limits based on the relationship between *u* and the original variable, and simply evaluate the antiderivative in terms of *u* values. Only the second method will be demonstrated, since the first is relatively self-explanatory.

EXAMPLE 6.27

Evaluate $\int_0^2 \dfrac{6x^2}{\sqrt{x^3+1}}\,dx$ by using u-substitution.

SOLUTION

Let $u = x^3 + 1$, then $du = 3x^2 dx$, so $2du = 6x^2 dx$.

Also when $x = 0$, $u = 0^3 + 1 = 1$. When $x = 2$, $u = 2^3 + 1 = 9$.

$$\int_0^2 \frac{6x^2}{\sqrt{x^3+1}}\,dx = \int_{u=1}^{u=9} u^{(-1/2)} \cdot 2\,du$$

$$= 4u^{(1/2)}\Big|_1^9$$

$$= 4\left[\sqrt{9} - \sqrt{1}\right]$$

$$= 8$$

EXAMPLE 6.28

Under the u-substitution $u = \cos(\pi\sqrt{x})$, what is the new form of the integral $\int_1^{25} \dfrac{\cos^2(\pi\sqrt{x})\sin(\pi\sqrt{x})}{\sqrt{x}}\,dx$?

SOLUTION

If $u = \cos(\pi\sqrt{x})$, then

$$du = -\sin(\pi\sqrt{x}) \cdot \frac{d}{dx}(\pi\sqrt{x})$$

$$= -\sin(\pi\sqrt{x}) \cdot \frac{\pi}{2\sqrt{x}}\,dx$$

So $\dfrac{-2}{\pi}\,du = \dfrac{\sin(\pi\sqrt{x})}{\sqrt{x}}\,dx$.

If $u = \cos(\pi\sqrt{x})$, when $x = 1$, then $u = \cos(\pi) = -1$.

If $u = \cos(\pi\sqrt{x})$, when $x = 25$, then $u = \cos(5\pi) = -1$

Substituting,

$$\int_{1}^{25} \frac{\cos^2(\pi\sqrt{x})\sin(\pi\sqrt{x})}{\sqrt{x}}\,dx = \int_{u=-1}^{u=-1} u^2 \cdot \frac{-2}{\pi}\,du$$

Notice that the value would be 0, since the limits are both -1.

6.7   EXERCISES

1. $\int 6\tan(2x)\,dx$

2. $\int \dfrac{dx}{\sqrt[3]{(5x+1)^2}}$

3. Antidifferentiate $\displaystyle\int \frac{\cos^3(x)}{1-\sin(x)}\,dx$ by rewriting the integral in a different form.

4. Given $f(x) = \dfrac{12}{1+x}$. Estimate $\displaystyle\int_0^3 f(x)\,dx$ using three right-oriented rectangles of equal width.

5. With equal subdivisions, use four trapezoids to estimate the value of $\displaystyle\int_1^5 (x^2-3)\,dx$; then find its actual value by integration.

6. If a function is nonnegative, does the concavity of a graph affect whether LRAM is an overestimate or underestimate of the area between a function and the x-axis? Explain.

7.

Table 6.4

Time (Seconds)	1	2	3	4	5	6	7
Velocity (Feet per second)	2	30	45	55	64	70	75

A vehicle moved with continuously increasing velocity, which was recorded at the times listed in Table 6.4. Use a midpoint Riemann sum with three equal subdivisions to estimate the distance traveled in the time interval [1, 7] seconds.

8. Write the definite integral equivalent of the Riemann sum,

$$\lim_{n \to \infty} \sum_{i=1}^{n} \frac{2x_i}{\sqrt{(x_i)^2 - 5}} \, \Delta x \text{ on the domain } [4, 8].$$

9. The graph of $h(x)$ given in Figure 6.14 consists of two segments with endpoints at $(-5, -2)$, $(-1, 0)$ and $(0, 2)$, as well as a quarter circle centered at $(0, 0)$ with a radius of 2 units.

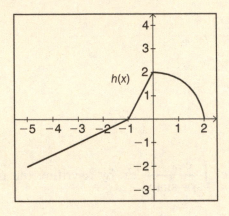

Figure 6.14

Use geometric methods to evaluate $\int_{2}^{-5} h(x) \, dx$.

10. If $\int_{1}^{3} f(x) \, dx = 8$, determine the value of $\int_{2}^{6} \left[3 + f\left(\frac{x}{2}\right) \right] dx$.

11. Let $g(x)$ be an odd function that is continuous for all real numbers, and let a, b, and c be constants.

If $\int_{a}^{b} g(x) \, dx = 12$ and $\int_{c}^{b} g(x) \, dx = -9$, find $\int_{c}^{a} \frac{g(-x)}{3} \, dx$.

12. Find the instantaneous rate of change of $k(x)$ if $k(x) = \int_{-6}^{x} [t \cdot \tan^{-1}(t^3) \, dt]$.

13. If $h(x) = \int_{\sin(x)}^{5} t^3 e^{(t^2)} \, dt$, find $h'(x)$.

14. If $f(x) = \int_{2}^{x^3} \ln(t + 4) \, dt$, is $f(x)$ increasing or decreasing when $x = -1$.

15. The graph of g is given in Figure 6.15. g is piecewise and consists of a semicircle of radius 2 units centered at $(-2, 1)$, a segment from $(0, 1)$ to $(2, -1)$ and a translated cubic function.

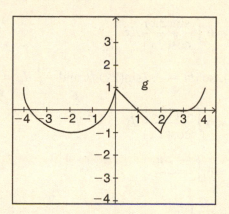

Figure 6.15

If $h(x) = \int_1^x g(t)\, dt$, determine the x-values in the interval $(-4, 4)$ where $h(x)$ has inflection points.

16. Evaluate $\displaystyle\int_0^{1/2} \frac{\sin^{-1}(x)}{\sqrt{1-x^2}}\, dx.$

17. Evaluate $\displaystyle\int_1^4 \frac{dx}{x+\sqrt{x}}.$

18. Evaluate $\displaystyle\int_1^2 x \cdot e^{(x^2-1)}\, dx.$

19. Evaluate $\displaystyle\int_0^1 (x^2-1)^2\, dx.$

20. Evaluate $\displaystyle\int_0^2 \frac{3\,dx}{4+x^2}.$

6.8 SOLUTIONS TO EXERCISES

1. $-3 \ln |\cos(2x)| + C$

   $$\int 6 \tan(2x)\, dx = 6 \int \frac{\sin(2x)}{\cos(2x)}\, dx$$

   Let $u = \cos(2x)$, so $du = -2\sin(2x)dx$ and $\frac{-1}{2} du = \sin(2x)dx$.

   $$\int 6 \tan(2x)\, dx = 6 \int \frac{\sin(2x)}{\cos(2x)}\, dx$$

   $$= 6 \int \frac{\left(-\frac{1}{2}\right)du}{u}$$

   $$= -3 \int \frac{du}{u}$$

   $$= -3 \ln |u| + C$$

   $$= -3 \ln |\cos(2x)| + C$$

2. $\frac{3}{5} \sqrt[3]{5x+1} + C$

   Rewrite so $\int \frac{dx}{\sqrt[3]{(5x+1)^2}} = \int (5x+1)^{(-2/3)}\, dx$.

   Let $u = 5x + 1$, so $du = 5dx$ and $\frac{1}{5} du = dx$.

   $$\int \frac{dx}{\sqrt[3]{(5x+1)^2}} = \int (5x+1)^{(-2/3)}\, dx$$

   $$= \int u^{(-2/3)} \cdot \frac{1}{5}\, du$$

   $$= \frac{1}{5} \cdot 3u^{(1/3)} + C$$

   $$= \frac{3}{5} \sqrt[3]{5x+1} + C$$

3. $\dfrac{1}{2}[1+\sin(x)]^2 + C$

$$\int \frac{\cos^3(x)}{1-\sin(x)}\,dx = \int \frac{\cos^2(x)\cdot \cos(x)}{1-\sin(x)}\,dx$$

$$= \int \frac{[1-\sin^2(x)]\cdot \cos(x)}{1-\sin(x)}\,dx$$

$$= \int \frac{[1-\sin(x)]\cdot [1+\sin(x)]\cdot \cos(x)}{1-\sin(x)}\,dx$$

$$= \int [1-\sin(x)]\cdot \cos(x)\,dx$$

Let $u = 1 + \sin(x)$, so $du = \cos(x)dx$

$$\int \frac{\cos^3(x)}{1-\sin(x)}\,dx = \int [1-\sin(x)]\cdot \cos(x)\,dx$$

$$= \int u\,du$$

$$= \frac{1}{2}u^2 + C$$

$$= \frac{1}{2}[1+\sin(x)]^2 + C$$

4. 13

$$\Delta x = \frac{3-0}{3} = 1$$

$$\int_0^3 \frac{12}{1+x}\,dx \approx f(1)\cdot 1 + f(2)\cdot 1 + f(3)\cdot 1$$

$$= \frac{12}{1+1} + \frac{12}{1+2} + \frac{12}{1+3}$$

$$= 13$$

5. Trapezoids: 30 Actual: $29\frac{1}{3}$

    Using trapezoids: $h = \Delta x = \dfrac{5-1}{4} = 1$

    $$\int_1^5 (x^2 - 3)\,dx \approx \frac{1}{2}[f(1) + 2f(2) + 2f(3) + 2f(4) + f(5)]$$

    $$= \frac{1}{2}[-2 + 2(1) + 2(6) + 2(13) + 22]$$

    $$= 30$$

    Actual value:

    $$\int_1^5 (x^2 - 3)\,dx = \frac{1}{3}x^3 - 3x \Big|_1^5$$

    $$= \left(\frac{1}{3}(125) - 3(5)\right) - \left(\frac{1}{3}(1) - 3(1)\right)$$

    $$= \frac{125}{3} - 15 - \frac{1}{3} + 3$$

    $$= 29\frac{1}{3}$$

6. Concavity does not affect whether LRAM is an overestimate or under-estimate. No matter whether a function is concave up or down, if it is increasing, the left endpoint of each subinterval will be lower than the right endpoint, the rectangles will lie below the function, and LRAM will be an underestimate. If the function is decreasing, LRAM will be an overestimate. Remember that concavity of a function is independent of whether it is increasing or decreasing.

7. 310 feet

    The intervals for the subdivisions are [1, 3], [3, 5], and [5, 7] so $\Delta x = 2$.

    The height of each rectangle is based on velocity at the midpoint of each interval.

    Distance $\approx 2[v(2) + v(4) + v(6)]$

    $\qquad = 2[30 + 55 + 70]$

    $\qquad = 310$ feet

8. $\displaystyle\int_4^8 \frac{2x}{\sqrt{x^2-5}}\, dx$

$\displaystyle\lim_{n\to\infty}\sum_{i=1}^{n} f(x_i)\Delta x = \int f(x)\, dx$, so by comparison $\displaystyle\lim_{n\to\infty}\sum_{i=1}^{n} \frac{2x_i}{\sqrt{(x_i)^2-5}}\Delta x =$

$\displaystyle\int_4^8 \frac{2x}{\sqrt{x^2-5}}\, dx$. The limits come from the endpoints of the domain $[4, 8]$.

9. $3-\pi$

Notice that integration is from 2 to -5, so Δx is negative. This means regions above the x-axis come out negative and regions below the x-axis come out positive.

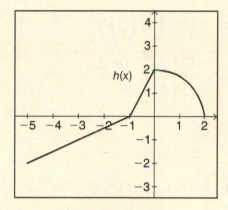

Figure 6.16

On $[0, 2]$, the quarter circle is $\dfrac{1}{4}\pi\cdot 2^2 = \pi$.

On $[-1, 0]$, the triangle is $\dfrac{1}{2}\cdot 1\cdot 2 = 1$.

On $[-5, -1]$, the triangle is $\dfrac{1}{2}\cdot 2\cdot 4 = 4$.

$\displaystyle\int_2^{-5} h(x)\, dx = -\pi + -1 + 4$

$\qquad\qquad = 3-\pi$

10. 28

$$\int_2^6 \left[3 + f\left(\frac{x}{2}\right) \right] dx = \int_2^6 3 \, dx + \int_2^6 f\left(\frac{x}{2}\right) dx$$

$$= 12 + \int_2^6 f\left(\frac{x}{2}\right) dx$$

On the remaining integral, use u-substitution, and equate it with the given integral.

Let $u = \dfrac{x}{2}$, so $du = \dfrac{1}{2} \, dx$ and $dx = 2 \, du$.

If $x = 2$, then $u = 1$. If $x = 6$, then $u = 3$

$$\int_2^6 \left[3 + f\left(\frac{x}{2}\right) \right] dx = 12 + \int_2^6 f\left(\frac{x}{2}\right) dx$$

$$= 12 + 2 \int_1^3 f(u) \, du$$

$$= 12 + 2 \cdot 8$$

$$= 28$$

11. 7

$$\int_a^c g(x) \, dx = \int_a^b g(x) \, dx + \int_b^c g(x) \, dx$$

$$= \int_a^b g(x) \, dx - \int_c^b g(x) \, dx$$

$$= 12 - (-9)$$

$$= 21$$

Since $g(x)$ is odd, $g(-x) = -g(x)$, so

$$\int_c^a \frac{g(-x)}{3} \, dx = \int_c^a \frac{-g(x)}{3} \, dx.$$

Also, by properties of definite integrals,

$$\int_c^a \frac{-g(x)}{3} \, dx = \int_a^c \frac{g(x)}{3} \, dx$$

$$\int_c^a \frac{g(-x)}{3} \, dx = \frac{1}{3} \int_a^c g(x) \, dx = 7$$

12. $k'(x) = x \cdot \tan^{-1}(x^3)$

The instantaneous rate of change is the derivative. By the Fundamental Theorem of Calculus, $\dfrac{d}{dx} \displaystyle\int_{-6}^{x} [t \cdot \tan^{-1}(t^3)]\, dt = x \cdot \tan^{-1}(x^3)$.

13. $h'(x) = -\sin^3(x) \cdot e^{[\sin^2(x)]} \cdot \cos(x)$

By the properties of definite integrals,

$$y = \int_{\sin(x)}^{5} t^3 e^{(t^2)}\, dt = -\int_{5}^{\sin(x)} t^3 e^{(t^2)}\, dt$$

Let $u = \sin(x)$ and $y = -\displaystyle\int_{5}^{u} t^3 e^{(t^2)}\, dt$.

$$\frac{dy}{dx} = \frac{dy}{du} \cdot \frac{du}{dx}$$

$$= [-u^3 e^{(u^2)}] \cdot \cos(x)$$

$$= -\sin^3(x) \cdot e^{[\sin^2(x)]} \cdot \cos(x)$$

14. Increasing

Determine if $f'(-1)$ is positive or negative.

If $f(x) = \displaystyle\int_{2}^{x^3} \ln(t+4)\, dt$, then by the Fundamental Theorem for composite functions,

$f'(x) = \ln(x^3 + 4) \cdot d(x^3)$

$\qquad = 3x^2 \ln(x^3 + 4)$

$f'(-1) = 3\ln(3) > 0$ so f is increasing at $x = -1$.

15. $h(x)$ has inflection points at $x = -2$, $x = 0$, and $x = 2$. Inflection points on h occur where $h''(x)$ changes sign. Since $h(x) = \displaystyle\int_{1}^{x} g(t)\, dt$, by the Fundamental Theorem, $h'(x) = g(x)$, and $h''(x) = g'(x)$. $h''(x) = g'(x)$ changes signs where g goes from increasing to decreasing or decreasing to increasing, at $x = -2$, $x = 0$, and $x = 2$.

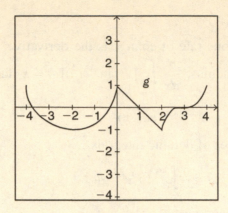

Figure 6.17

16. $\dfrac{\pi^2}{72}$

Let $u = \sin^{-1}(x)$ so $du = \dfrac{dx}{\sqrt{1-x^2}}$.

When $x = 0$, $u = \sin^{-1}(0) = 0$.

When $x = \dfrac{1}{2}$, $u = \sin^{-1}\left(\dfrac{1}{2}\right) = \dfrac{\pi}{6}$.

$$\int_0^{1/2} \dfrac{\sin^{-1}(x)}{\sqrt{1-x^2}}\, dx = \int_0^{\pi/6} u\, du$$

$$= \dfrac{u^2}{2}\Big|_0^{\pi/6}$$

$$= \dfrac{\pi^2}{72}$$

17. $2[\ln(3) - \ln(2)]$

To change the form, factor $\sqrt{x}$ out of the denominator.

$$\int_1^4 \dfrac{dx}{x+\sqrt{x}} = \int_1^4 \dfrac{dx}{\sqrt{x}(\sqrt{x}+1)}$$

Let $u = \sqrt{x}+1$, so $du = \dfrac{1}{2\sqrt{x}} \cdot dx$ and $\dfrac{dx}{\sqrt{x}} = 2\,du$.

When $x = 1$, $u = 2$.

When $x = 4$, $u = 3$.

$$\int_1^4 \frac{dx}{\sqrt{x}(\sqrt{x}+1)} = \int_2^3 \frac{2\,du}{u}$$

$$= 2\ln|u|\Big|_2^3$$

$$= 2[\ln(3) - \ln(2)]$$

18. $\dfrac{e^3 - 1}{2}$

Let $u = x^2 - 1$, so $du = 2x\,dx$, and $x\,dx = \dfrac{1}{2}\,du$.

When $x = 1$, $u = 0$.

When $x = 2$, $u = 3$.

$$\int_1^2 x \cdot e^{(x^2-1)}\,dx = \int_0^3 e^u \cdot \frac{1}{2}\,du$$

$$= \frac{1}{2}\int_0^3 e^u\,du$$

$$= \frac{e^u}{2}\Big|_0^3$$

$$= \frac{e^3 - 1}{2}$$

19. $\dfrac{8}{15}$

Change the form by expanding the binomial square, and integrate term by term.

$$\int_0^1 (x^2 - 1)^2\,dx = \int_0^1 (x^4 - 2x^2 + 1)\,dx$$

$$= \frac{x^5}{5} - \frac{2}{3}x^3 + x\Big|_0^1$$

$$= \frac{1}{5} - \frac{2}{3} + 1$$

$$= \frac{8}{15}$$

20. $\dfrac{3\pi}{8}$

Factor a 4 out of the denominator.

$$\int_0^2 \frac{3dx}{4+x^2} = \int_0^2 \frac{3dx}{4\left(1+\frac{x^2}{4}\right)}$$

Let $u^2 = \dfrac{x^2}{2}$, so $u = \dfrac{x}{2}$, and $du = \dfrac{1}{2}dx$ or $2du = dx$.

When $x = 0$, $u = 0$.

When $x = 2$, $u = 1$.

$$\int_0^2 \frac{3\,dx}{4+x^2} = \int_0^2 \frac{3\,dx}{4\left(1+\frac{x^2}{4}\right)}$$

$$= \frac{3}{4} \int_0^1 \frac{2\,du}{1+u^2}$$

$$= \frac{3}{2} \tan^{-1} u \Big|_0^1$$

$$-\frac{3}{2}[\tan^{-1}(1) - \tan^{-1}(0)]$$

$$= \frac{3}{2} \cdot \frac{\pi}{4}$$

$$= \frac{3\pi}{8}$$

CHAPTER 7

Applications of Integrals

APPLICATIONS OF INTEGRALS

7.1 INTRODUCTION

With antidifferentiation skills intact, it is now time to explore the multitude of applications of integrals. Antidifferentiation enables mathematicians, scientists, engineers, economists, and others to "work backwards" from knowledge about changes in quantities to create models of population growth, to determine position functions from knowledge about velocity, to compute average energy use over time, to find marginal cost, and to perform countless other applications.

7.2 DIFFERENTIAL EQUATIONS

SEPARABLE DIFFERENTIAL EQUATIONS

A differential equation is any equation that relates the instantaneous rate of change of one variable with respect to another to some mathematical combination of the independent and dependent variable. That may sound confusing, but let's look at a basic example such as $\frac{dy}{dx} = \frac{x+1}{y}$. The left side of the equation is the "instantaneous rate of change of one variable with respect to another." The right side of the equation is "some mathematical combination of the independent and dependent variable." In many cases, the dependent variable may not appear at all in the expression, for example, $\frac{dy}{dx} = 2x + 3$. Differential equations are one of the building blocks of calculus. All differential equations encountered in the first year of calculus are separable. This means, if necessary, all y terms can be grouped with the differential dy and all x terms can be grouped with the differential dx. Of course, a differential equation may be a function of variables other than x and y, but they are all handled the same way: separate the variables and antidifferentiate each side of the equation with respect to the appropriate variable.

EXAMPLE 7.1

Solve the differential equation, $\dfrac{dy}{dx} = 3x + 5$.

SOLUTION

Separate the variables. $\dfrac{dy}{dx} = 3x + 5$ becomes $dy = (3x + 5)dx$.

Now integrate both sides of the equation.

$$\int dy = \int (3x + 5)\, dx$$

$$y + C_1 = \frac{3}{2}x^2 + 5x + C_2$$

Since the constants are arbitrary, they may be combined into one constant on either side of the equation. In general, when possible, isolate the dependent variable.

$$y = \frac{3}{2}x^2 + 5x + C$$

Notice that the solution to the differential equation is a family of functions differing by a constant. The solution represents all functions for which $\dfrac{dy}{dx} = 3x + 5$.

EXAMPLE 7.2

Solve $\dfrac{ds}{dt} = s \cdot (2t + 1)$

SOLUTION

$$\frac{ds}{s} = (2t+1)dt$$

$$\int \frac{ds}{dt} = \int (2t+1)\,dt$$

$$\ln|s| = t^2 + t + C$$

$$|s| = e^{(t^2+t+C)} \qquad \text{(Using } \ln(a) = b \Rightarrow e^b = a.)$$

$$|s| = e^C e^{(t^2+t)} \qquad \text{(Using } e^{(a+b)} = e^a \cdot e^b.)$$

$$s = Ae^{(t^2+t)}, \text{ where } A = \pm e^C$$

DIFFERENTIAL EQUATIONS WITH INITIAL CONDITIONS

In the absence of any more information, finding the general family of functions as a solution to the differential equation is all that can be done. But all it takes is knowledge about one point on one of the functions in the family of solutions, and the infinite number of solutions can be narrowed to exactly one specific solution. Since they all differ by a constant, and all have the same derivative, none of the infinite number of solution curves has a single point in common. So knowing one specific point identifies the one and only function that contains that point, and from that, the constant can be determined. Knowledge of a point on the solution curve is called having an initial condition. The process then is just one or two steps beyond solving for the family of curves. After integration of the indefinite integral, substitute the known values of the independent and dependent variables and solve for the constant C.

EXAMPLE 7.3

The rate of change in the height, in feet per second, of a vertically ascending helium balloon is a function of time, $\frac{dh}{dt} = 2\sqrt{t}$. If the balloon was 8 feet off the ground when the holder let go, find a function for the height of the balloon at any time t seconds.

SOLUTION

$$\frac{dh}{dt} = 2\sqrt{t} \text{ so } dh = 2\sqrt{t}\,dt$$

$$\int dh = \int 2\sqrt{t}\,dt$$

$$h = \frac{4}{3}t^{\left(\frac{3}{2}\right)} + C$$

But at time $t = 0$ seconds (when the balloon was let go), its height was $h = 8$ feet.

$$8 = \frac{4}{3} \cdot 0^{\left(\frac{3}{2}\right)} + C \Rightarrow C = 8$$

So the function is:

$$h = \frac{4}{3}t^{\left(\frac{3}{2}\right)} + 8$$

With this information, the height of the balloon can be determined at any given time.

Even though the next example is somewhat contrived to work nicely without a calculator, the hope is that it sheds a bit of light on the possibility of economic modeling with differential equations.

EXAMPLE 7.4

Most small businesses expect to lose money for a certain initial amount of time, but eventually they have a break-even month in which income overtakes expenses. Let the instantaneous rate of change of percentage growth in income of a small business t months after the break-even month be defined by $\frac{dP}{dt} = \frac{1}{1+(t-1)^2}$. Also, suppose $P(1) = \frac{\pi}{4}$, which means one month after the break-even month, the percentage growth in income is about 0.785%. Find a model for percentage growth in income as a function of months beyond the break-even month.

SOLUTION

$$\frac{dP}{dt} = \frac{1}{1+(t-1)^2}, \text{ so } dP = \frac{1}{1+(t-1)^2} dt$$

$$\int dP = \int \frac{1}{1+(t-1)^2} dt$$

$$P = \tan^{-1}(t-1) + C$$

Using $P(1) = \dfrac{\pi}{4}$, $\dfrac{\pi}{4} = \tan^{-1}(1-1) + C$.

Therefore, $C = \dfrac{\pi}{4}$, so $P = \tan^{-1}(t-1) + \dfrac{\pi}{4}$.

POSITION, VELOCITY, AND ACCELERATION

Example 7.3 already alluded to the relationship between velocity and position, but this topic is significant enough to deserve special focus. The differential calculus portion of this text established that velocity is the derivative of position and acceleration is the derivative of velocity. It follows logically, then, that velocity is the antiderivative of acceleration, and position is the antiderivative of velocity. So if the velocity of a particle can be modeled with a function, and the position of that particle is known at any one moment, by solving a differential equation, it is possible to determine the position of the particle at any time.

EXAMPLE 7.5

A particle moves along the x-axis such that its velocity, in feet per second, at any time t seconds is $v(t) = t \cdot \cos(t^2)$. If it starts 4 feet from the origin, find a function, $s(t)$, for the position of the particle at any time t.

SOLUTION

$$v(t) = \frac{ds}{dt}, \text{ so } \frac{ds}{dt} = t \cdot \cos(t^2), \text{ and } ds = t \cdot \cos(t^2) dt$$

$$\int ds = \int t \cdot \cos(t^2) dt$$

$$s = \frac{1}{2} \sin(t^2) + C, \text{ and at time } t = 0, s = 4.$$

$$4 = \frac{1}{2}\sin(0) + C \Rightarrow C = 4, \text{ so } s(t) = \frac{1}{2}\sin(t^2) + 4.$$

Sometimes, more than one initial condition is necessary, as shown in Example 7.6.

EXAMPLE 7.6

Let $v(t)$ be velocity, and $s(t)$ be position. If $\frac{dv}{dt} = 2t$, $v(1) = 3$ and $s(1) = 4$, find position as a function of time.

SOLUTION

$\frac{dv}{dt} = 2t$, so $dv = 2t\,dt$.

$\int dv = \int 2t\,dt$, so $v = t^2 + C_1$.

Using $v(1) = 3$ leads to $3 = 1^2 + C_1 \Rightarrow C_1 = 2$.

$v = \frac{ds}{dt} = t^2 + 2$, so $ds = (t^2 + 2)dt$.

$\int ds = \int (t^2 + 2)\,dt$, so $s = \frac{1}{3}t^3 + 2t + C_2$.

Using $s(1) = 4$ leads to $4 = \frac{1}{3}(1^3) + 2(1) + C_2 \Rightarrow C_2 = \frac{5}{3}$.

$$s = \frac{1}{3}t^3 + 2t + \frac{5}{3}$$

EXAMPLE 7.7

Near Earth's surface, ignoring air resistance, an object free falls with a constant acceleration of $-32\,\frac{\text{ft}}{\text{sec}^2}$. With what velocity will an object dropped from a height of 144 feet hit the ground?

SOLUTION

$\dfrac{dv}{dt} = -32$, so $dv = -32dt$

$\displaystyle\int dv = \int -32\,dt$

$v = -32t + C_1$

Since it was dropped and not thrown, at time $t = 0$, $v = 0$, which makes $C_1 = 0$.

Now $v = \dfrac{ds}{dt} = -32t\,dt$, so $ds = -32t\,dt$

$\displaystyle\int ds = \int -32t\,dt$

$s = -16t^2 + C_2$

Since at $t = 0$, $s = 144$ feet, $C_2 = 144$ and $s = -16t^2 + 144$.

When the object hits the ground, $s = 0$, so $0 = -16t^2 + 144 \Rightarrow t = 3$, and $v(3) = -32(3) = -96$ feet per second. The negative just indicates downward direction. The object strikes the ground with a speed of 96 feet per second, which is more than 60 miles per hour!

EXPONENTIAL GROWTH AND DECAY

A vast number of functions in science and economics behave exponentially and have a unique phenomenon regarding their rates of change. For an exponential function of the form, $y = a \cdot e^{kx}$, where a and k are constants, the ratio of the slope at any point of the function to the function value at that point is constant. A common way to state this is: "The rate of change of a function is proportional to the function." In symbols, it is written, $\dfrac{dy}{dt} = k \cdot y$, where y is a function of t and k is a constant sometimes called the relative growth rate. This holds true for unchecked growth of a population, radioactive decay of isotopes, velocity of an object coasting to a stop, capacitor discharge, and growth of investments under continuously compounded interest, to name a few. Solving the differential equation will show how an exponential function results.

If $\dfrac{dy}{dt} = k \cdot y$, then $\dfrac{dy}{y} = k \cdot dt$.

$$\int \frac{dy}{y} = \int k \cdot dt$$

$$\ln |y| = k \cdot t + C$$

$$|y| = e^{kc+C} = e^C e^{kt}$$

So $y = Ae^{kt}$ where $A = \pm\, e^c$. The sign and value of A is determined by the initial condition.

EXAMPLE 7.8

The instantaneous rate of growth of the population of a booming suburban town is 4% of the population at any given time. If the population this year is 130,000 people, find a model for the population as a function of years beyond this current year.

SOLUTION

$$\frac{dP}{dt} = 0.04P$$

$$\frac{dP}{P} = 0.04dt$$

$$\int \frac{dP}{P} = \int 0.04\, dt$$

$$\ln |P| = 0.04t + C$$

$$|P| = e^{0.04t+C}$$

$$P = Ae^{0.04t}$$

Currently, $t = 0$, and $P(0) = 130,000$.

So $130,000 = Ae^{0.04(0)} \Rightarrow 130,000 = A$

and $P = 130,000e^{0.04t}$

Of course, once the conditions that give rise to an exponential function are recognized, all the solution steps can be bypassed, and one can jump directly to the solution.

EXAMPLE 7.9

A capacitor is losing voltage at a rate proportional to the voltage passing over its terminals. Let $\dfrac{dV}{dt} = -0.02V$, where t is time in seconds. Find the exact amount of time for the capacitor to lose 60% of its initial charge.

SOLUTION

Even though the initial charge was not given, the question asks about a percentage of the original charge so let V_i be the initial charge.

We bypass the integration process and move right to the exponential model,

$V = V_i \cdot e^{-0.02t}$

If the capacitor loses 60%, then 40% remains.

So we solve $0.40V_i = V_i \cdot e^{-0.02t}$.

Divide both sides by V_i to get $0.4 = e^{-0.02t}$

$\ln(0.4) = \ln(e^{-0.02t})$
$\ln(0.4) = -0.02t$
$t = \dfrac{\ln(0.4)}{-0.02}$ seconds

This is the exact value, and with a calculator, we find it is about 45.8 seconds.

COMPOUNDED INTEREST

You may recall from a previous course that the formula for the value of an investment subject to compounded interest is $P = P_0\left(1 + \dfrac{r}{n}\right)^{(n \cdot t)}$ where P_0 is the initial investment, r is the annual interest rate, n is the number of times interest is compounded each year, and t is the number of years of the investment. To move to continuous compounding, the number of times interest is compounded each year must increase without bound. In essence, let $n \to \infty$. This is actually a limit that is calculated in the next calculus course, but the result is presented here without proof. $\lim_{n \to \infty} = P_0\left(1+\dfrac{r}{n}\right)^{(n \cdot t)} = P_0 e^{(r \cdot t)}$. Therefore, the formula for growth of an investment subject to continuously compounded interest is $P = P_0 e^{(r \cdot t)}$.

EXAMPLE 7.10

How long will it take to double an initial investment at 3% annual interest compounded continuously?

SOLUTION

Again, the initial amount does not matter since doubling any amount will take the same amount of time.

$2P_0 = P_0 e^{(0.03t)}$

$2 = e^{(0.03t)}$

$\ln(2) = \ln[e^{(0.03t)}] = 0.03t$

Then $t = \dfrac{\ln(2)}{0.03} \approx 23.1$ years.

DISTANCE AND DISPLACEMENT

Example 6.15 first raised the discussion about the difference between total and net area under a velocity function as it applies to total distance and displacement. With our increased skills, let's now take a brief second look at the topic and formalize the main ideas.

Displacement

For an object in linear motion, the definite integral of velocity on a given interval finds the net area between the velocity graph and the time axis. This is the displacement of an object. If $v(t)$ is the velocity of an object on the time interval $[a, b]$, then displacement $= \displaystyle\int_a^b v(t)\,dt$.

Total Distance

For an object in linear motion, the definite integral of the absolute value of velocity on a given interval finds the total area between the velocity graph and the time axis. This is the total distance traveled. If $v(t)$ is the velocity of an object on the time interval $[a, b]$, then total distance = $\int_a^b |v(t)| \, dt$.

Displacement is certainly the easier of the two quantities to find since it does not require determining where a velocity function is above or below the time axis. Just integrate the definite integral of velocity over the entire time interval and displacement results. The total distance formula can be a bit misleading because, without technology, there is no simple way to take the integral of the absolute value of a function. It must be determined on which intervals the velocity is positive and negative, and separate integrals must be written for each interval so that negative quantities can be treated as positive.

EXAMPLE 7.11

If the velocity of a particle is $v(t) = -t^2 + 2t$ for the time interval $[0, 3]$, determine displacement and total distance traveled during that time.

SOLUTION

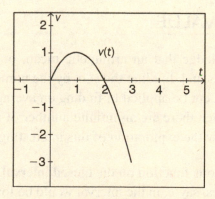

Figure 7.1

$$\int_0^3 (-t^2 + 2t)\, dt = \frac{-1}{3} t^3 + t^2 \Big|_0^3$$

$$= \left(\frac{-1}{3} \cdot 27 + 3^2 \right) - (0 + 0) = 0$$

Displacement $= 0$

For total distance, negative velocity must be treated as positive. We see from the graph of $v(t)$ in Figure 7.1 that the velocity changes to a negative at $t = 2$. So

$$\int_0^2 (-t^2 + 2t)\, dt - \int_2^3 (-t^2 + 2t)\, dt$$

$$= \left(\frac{-1}{3} t^3 + t^2 \right) \Big|_0^2 - \left(\frac{-1}{3} t^3 + t^2 \right) \Big|_2^3$$

$$= \left(\frac{-8}{3} + 4 \right) - \left[\left(\frac{-27}{3} + 9 \right) - \left(\frac{-8}{3} + 4 \right) \right]$$

$$= \left(\frac{4}{3} \right) - \left[(0) - \left(\frac{4}{3} \right) \right]$$

$$= \frac{8}{3}$$

Total distance is $\dfrac{8}{3}$ units.

7.3 AVERAGE VALUE

It is common knowledge that an arithmetic mean, or average, is found by summing a set of values and dividing the sum by the number of elements in the set. But could this concept be applied to finding an average value of a function on a closed interval when there are an infinite number of points in the interval? Once again, limits allow the exploration of this interesting idea.

Let $f(x)$ be a continuous function on the closed interval $[a, b]$. The average of a finite number of values, say n, in the interval would be found by

$$y_{avg} = \frac{f(x_1) + f(x_2) + f(x_3) + \ldots + f(x_{n-1}) + f(x_n)}{n}$$

For convenience, the x-values chosen to generate each y-value could be evenly distributed across the interval. This would generate a distance between each x, which is just Δx from Riemann sums. As usual, $\Delta x = \dfrac{b-a}{n}$ which makes $n = \dfrac{b-a}{\Delta x}$. By substitution,

$$y_{\text{avg}} = \frac{f(x_1)+f(x_2)+f(x_3)+\ldots+f(x_{n-1})+f(x_n)}{\frac{b-a}{\Delta x}}, \text{ so}$$

$$y_{\text{avg}} = \frac{f(x_1)+f(x_2)+f(x_3)+\ldots+f(x_{n-1})+f(x_n)}{b-a} \cdot \Delta x.$$

To get a true average, the number of function values used must approach infinity, so a limit is used.

$$y_{\text{avg}} = \lim_{n\to\infty} \frac{f(x_1)+f(x_2)+f(x_3)+\ldots+f(x_{n-1})+f(x_n)}{b-a} \cdot \Delta x$$

$$= \frac{1}{b-a} \lim_{n\to\infty} \sum_{i=1}^{n} f(x_i)\Delta x$$

Does this result look familiar? It should!

Average Value

> If $f(x)$ is continuous on the closed interval $[a, b]$, then the average value of f is $f_{avg} = \dfrac{1}{(b-a)} \displaystyle\int_a^b f(x)\,dx$.

Even though it is interesting and informative to think of this as the average of all function values in the interval, it is perhaps more enlightening to think of it geometrically.

A simple rearrangement of the equation gives $f_{\text{avg}} \cdot (b-a) = \displaystyle\int_a^b f(x)\,dx$. For simplicity, consider $f(x)$ to be positive over the entire interval $[a, b]$. The integral on the right side of the equation then finds the total area under the graph of $f(x)$ and above the x-axis. On the left side of the equation, $(b-a)$ is the width of the interval, and f_{avg} is a height above the x-axis. So the equation is really height times width equals area. f_{avg} is the height of the single rectangle with width $(b - a)$ which contains the same amount of area as the area under the curve.

Figure 7.2 illustrates this fascinating idea.

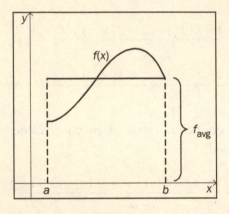

Figure 7.2

The function was chosen to be positive to simplify the introduction of the idea, but there is absolutely no reason f_{avg} cannot be computed for any continuous function on a given interval. In fact, it is certainly possible for an average value to be negative or even zero.

The geometric interpretation of average value lends itself to a logical conclusion, which is the integral calculus version of the Mean Value Theorem. The height of the single rectangle will fall somewhere between the maximum value of the function, and the minimum value of the function on the interval $[a, b]$. As a result, the upper edge of the rectangle must intersect the function somewhere between the endpoints of the interval $[a, b]$. This guarantees there will be at least one value in (a, b) that generates f_{avg}.

Mean Value Theorem for Definite Integrals

If a function $f(x)$ is continuous on an interval $[a, b]$, then there exists at least one value $x = c$ in (a, b) such that

$$f(c) = \frac{1}{(b-a)} \int_a^b f(x)\,dx.$$

EXAMPLE 7.12

Find the average value of $g(x) = \dfrac{2-x}{x}$ on the interval $[1, 5]$.

SOLUTION

$$\frac{1}{5-1} \int_1^5 \left(\frac{2-x}{x}\right) dx = \frac{1}{4} \int_1^5 \left(\frac{2}{x} - 1\right) dx$$

$$= \frac{1}{4} [2 \ln(x) - x] \Big|_1^5$$

$$= \frac{1}{4} ([2 \ln(5) - 5] - [2 \ln(1) - 1])$$

$$= \frac{1}{4} [2 \ln(5) - 5 - 0 + 1]$$

$$= \frac{1}{2} \ln(5) - 1$$

In physics, force times distance equals work. When force is constant, calculating work is simple. To raise a 3-pound book 2 feet takes 6 foot-pounds of work. But when force is constantly varying, calculus is useful to find the work done. Any given force can be applied only over an infinitely small distance and then must change. As a result, an infinite number of products of force times distance must be summed up. Summing up an infinite number of products ought to sound like a Riemann sum whose limit is an integral.

EXAMPLE 7.13

To stretch a spring x meters beyond its resting point requires a force of $6x$ Newtons. Find the average amount of work done in stretching the spring from equilibrium to 5 meters beyond equilibrium.

SOLUTION

Force applied at any point from 0 to 5 meters is $f = 6x_i$ and is applied only over a small change in x, Δx. Work $= \displaystyle\lim_{n \to \infty} \sum_{i=1}^{n} f(x_i) \cdot \Delta x$

Average work $= \dfrac{1}{5-0} \displaystyle\int_0^5 6x\, dx = \dfrac{1}{5} \cdot 3x^2 \Big|_0^5$

$\dfrac{3}{5} x^2 \Big|_0^5 = \dfrac{3}{5} (25 - 0) = 15$ N - m

EXAMPLE 7.14

On the interval [1, 5], which of the following choices is the best estimate of the average value of the function $h(x)$ shown in Figure 7.3?

(A) 1

(B) 1.5

(C) 2.5

(D) 4

(E) 6

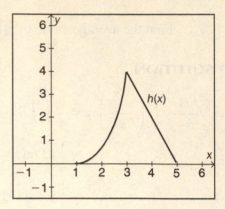

Figure 7.3

SOLUTION

The area under $h(x)$ on [3, 5] is $\frac{1}{2} \cdot 2 \cdot 4 = 4$ units. The area under $h(x)$ on [1, 3] is about 2 units, and the width $b - a = 4$ units. Dividing the area 6 units by 4 is an average of 1.5, which is choice B.

EXAMPLE 7.15

Find the value of c that satisfies the Mean Value Theorem for Integrals for the function $g(x) = 3x^2 + 1$ on the interval [0, 2].

SOLUTION

$$\frac{1}{2-0} \int_0^2 (3x^2 + 1)\, dx = \frac{1}{2}(x^3 + x)\Big|_0^2$$

$$= \frac{1}{2}(2^3 + 2) = 5$$

$$g(c) = \frac{1}{2-0} \int_0^2 (3x^2 + 1)\, dx \Rightarrow 3c^2 + 1 = 5$$

$c^2 = \frac{4}{3}$ so $c = \frac{2}{\sqrt{3}} = \frac{2\sqrt{3}}{3}$. The negative root is omitted since it is not in [0, 2].

7.4   AREA

AREA BETWEEN TWO CURVES

Much of the work to this point has dealt with the net or total area between a single function and the x-axis. There are times, though, where several functions exist, and the area between those functions is desired. If your understanding of area from Riemann sums is strong, then the concepts presented here should go smoothly. To determine the area between two curves on an interval $[a, b]$, instead of drawing rectangles from each curve to the x-axis, rectangles are simply drawn from function to function. The differences between the two function values, larger minus smaller, will then determine the height of each rectangle. Additionally, if the smaller function value is always subtracted from the larger value, it does not matter whether the functions lie above or below the x-axis. Again, as the width of each rectangle approaches 0, it also does not matter which x_i in each subinterval is used to determine function values being subtracted.

Figure 7.4 illustrates the idea with a few rectangles.

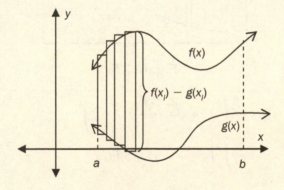

Figure 7.4

Area Between Two Functions

If functions $f(x)$ and $g(x)$ are each continuous on an interval $[a, b]$, and $f(x) \geq g(x)$ over the entire interval, the area between f and g is $\lim_{n \to \infty} \sum_{i=1}^{n} [f(x_i) - g(x_i)] \cdot \Delta x = \int_a^b [f(x) - g(x)] \, dx$.

If the functions happen to cross each other inside the interval of integration, two integrals may be required with the order of subtraction of the functions changing from one to the other.

EXAMPLE 7.16

Find the area between $g(x) = 3 + \sqrt{x}$ and $f(x) = \dfrac{1}{4}x^2 - 2$ on $[0, 4]$.

SOLUTION

Figure 7.5 shows a sketch of the two curves and the area between them.

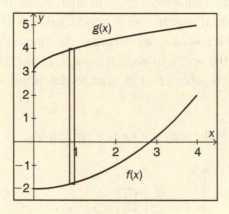

Figure 7.5

$$\int_0^4 [g(x) - f(x)]\, dx = \int_0^4 (3 + \sqrt{x}) - \left(\frac{1}{4}x^2 - 2\right)$$

$$= \int_0^4 \left(\sqrt{x} - \frac{1}{4}x^2 + 5\right) dx$$

$$= \frac{2}{3}x^{\left(\frac{3}{2}\right)} - \frac{1}{12}x^3 + 5x \Big|_0^4$$

$$= \frac{2}{3}4^{\left(\frac{3}{2}\right)} - \frac{1}{12}(4)^3 + 5(4)$$

$$= \frac{16}{3} - \frac{16}{3} + 20$$

$$= 20$$

EXAMPLE 7.17

The graphs of $g(x)$ and $h(x)$ are shown in Figure 7.6. Express the area between the curves on $[a, b]$ three different ways without using absolute value symbols.

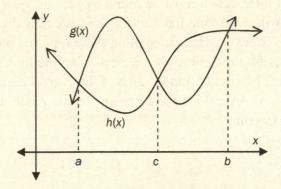

Figure 7.6

SOLUTION

The first expression always uses the larger function minus the smaller.

$$\int_a^c [g(x) - h(x)]\,dx + \int_c^b [h(x) - g(x)]\,dx$$

The second expression keeps the order of subtraction the same, but subtracts the second integral, which would have a negative value.

$$\int_a^c [g(x) - h(x)]\,dx - \int_c^b [g(x) - h(x)]\,dx$$

The third expression maintains the order of subtraction, but reverses the order of the limits to make the negative quantity positive.

$$\int_a^c [g(x) - h(x)]\,dx + \int_b^c [g(x) - h(x)]\,dx$$

Can you think of more ways?

BOUNDED REGIONS

Sometimes, the interval over which area is being found is not given. Instead, the phrase, "bounded by" is used and followed by given equations that create an enclosed region between their points of intersection. In cases such as this, the points of intersection must be found by solving systems of equations, and those points will determine the limits on the integrals used. There are two common mistakes to avoid when attempting these types of problems. First, be careful to correctly find the points of intersection so the limits on the integrals are correct. Second, be sure to find the area of the correct region. Often, it may appear that there are several enclosed regions in the plane and those "extra" regions can be distracting.

EXAMPLE 7.18

Find the area of the region enclosed by $f(x) = x^2$ and the line $g(x) = x + 2$.

SOLUTION

$$f(x) = g(x) \Rightarrow x^2 = x + 2 \text{ so } x^2 - x - 2 = 0$$

Factoring, $(x - 2)(x + 1) = 0 \Rightarrow x = 2$ or $x = -1$

The integrand is the line minus the parabola, because the line lies above the parabola.

$$
\begin{aligned}
\int_{-1}^{2} (x + 2 - x^2)\, dx &= \frac{1}{2} x^2 + 2x - \frac{1}{3} x^3 \bigg|_{-1}^{2} \\
&= \left(2 + 4 - \frac{8}{3}\right) - \left(\frac{1}{2} - 2 + \frac{1}{3}\right) \\
&= \frac{9}{2}
\end{aligned}
$$

EXAMPLE 7.19

Find the first quadrant area bounded by $f(x) = 2\sin(x)$, $g(x) = 2$ and the y-axis.

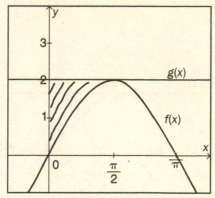

Figure 7.7

SOLUTION

First, we draw the graph (Figure 7.7). Even though the line intersects the sine curve at its peaks an infinite number of times, a three-sided region is described. Only the first quadrant, enclosed region adjacent to the y-axis, fits the description.

$$\int_{-1}^{(\pi/2)} [2 - 2\sin(x)]\,dx = [2x + 2\cos(x)]\Big|_0^{(\pi/2)}$$

$$= 2 \cdot \frac{\pi}{2} + 2\cos\left(\frac{\pi}{2}\right) - [2(0) + 2\cos(0)]$$

$$= \pi + 2(0) - 0 - 2(1)$$

$$= \pi - 2$$

SUMMING TWO REGIONS

One more variation on area between curves arises when the functions defining the heights of the rectangles change in the interior of the interval. For one portion of the interval, $f(x) - g(x)$ may be the height of the rectangles, but then at some point c in the interval it may change to $h(x) - g(x)$ defining the heights. A quick sketch of the region is always recommended so a situation such as this is not overlooked.

EXAMPLE 7.20

Set up an integral or integrals to express the area enclosed by the graphs of $f(x) = x^3 + 1$, $g(x) = -x + 3$, and $h(x) = \dfrac{1}{3}x + \dfrac{1}{3}$. Do not integrate or evaluate.

SOLUTION

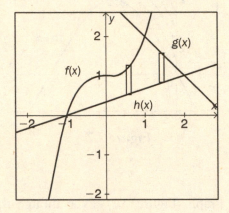

Figure 7.8

Figure 7.8 shows a sketch of the three functions. Notice on the interval $[-1, 1]$, $f(x) - h(x)$ determines the heights of the rectangles, but g and f intersect at the point $(1, 2)$ so on the interval $[1, 2]$, $g(x) - h(x)$ now determines the heights of the rectangles.

$$\text{Area} = \int_{-1}^{1}\left[(x^3+1)-\left(\frac{1}{3}x+\frac{1}{3}\right)\right] dx + \int_{1}^{2}\left[(-x+3)-\left(\frac{1}{3}x+\frac{1}{3}\right)\right] dx$$

It should be apparent that success with this topic requires a solid grasp of prerequisite skills such as knowledge of the more commonly used functions, an ability to quickly sketch functions and their transformations, and mastery of solving systems of equations to find the points of intersection, which determine the limits of the integral expressions.

7.5   EXERCISES

1. Solve the differential equation $\dfrac{dp}{dr} = \dfrac{1}{pr + 3p}$.

2. Find the y as a function of x if $\dfrac{dy}{dx} = \dfrac{4x}{\sqrt{x^2 - 5}}$ and $y(3) = 6$.

3. An object moving along the x-axis has a velocity of $v(t) = 4 - t^2$. If at $t = 1$ second its position is $x = 5$, find its position at $t = 3$ seconds.

4. The graph of the acceleration of an object is shown in Figure 7.9. If $v(0) = -1$, will $v(6)$ be positive or negative? Explain.

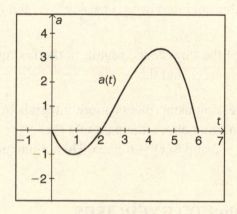

Figure 7.9

5. An object is thrown upward with a velocity of 64 feet per second from a height of 80 feet. How long will it take to hit the ground?

6. A population of bacteria grow at a rate proportional to the size of the population. If the culture starts with 2 bacteria on day 0, and has 64 bacteria on day 4, find an equation to model its growth.

7. The velocity of an object in motion is given by $v(t) = (t + 1)^2 - a$. Find the value of a if the displacement of the object in the first two seconds is $\dfrac{2}{3}$.

8. The velocity of a particle moving along a line is $v(t) = t^2 - 3t + 2$ feet per second. Find the total distance traveled in the interval $[0, 2]$ seconds.

9. Find the average value of $f(x) = 2x + \cos(x)$ on the interval $\left[0, \dfrac{3\pi}{2}\right]$.

10. What is the average length of a chord in a circle of radius 6?

11. If the average value of a function on a closed interval is negative, does this mean the function has to be negative over more than half the interval? Explain.

12. Find the area between $f(x) = \dfrac{1}{2}x + 4$ and $h(x) = \dfrac{2}{x+1}$ on $[0, 4]$.

13. Find the total area enclosed by $g(x) = \dfrac{1}{4}x^3 - x$ and $h(x) = 3x$

14. Find the area of the three-sided region in the first quadrant bounded by $y = \sin(x)$, $y = \cos(x)$, and the y-axis.

15. Set up an expression using one or more integrals to find the area of the bounded region lying above the parabola $f(x) = x^2 - 4$, and below the lines $g(x) = x + 2$ and $h(x) = 6 - 3x$. Do not integrate or evaluate.

7.6 SOLUTIONS TO EXERCISES

1. $p = \pm\sqrt{2\ln(r+3) + 2C}$

$\dfrac{dp}{dr} = \dfrac{1}{pr + 3P} = \dfrac{1}{p(r+3)}$ so $p \cdot dp = \dfrac{dr}{r+3}$

$\displaystyle\int p \cdot dp = \int \dfrac{dr}{r+3}$

$\dfrac{1}{2}p^2 = \ln(r+3) + C$

$p^2 = 2\ln(r+3) + 2C \Rightarrow p = \pm\sqrt{2\ln(r+3) + 2C}$

2. $y = 4\sqrt{x^2 - 5} - 2$

$\dfrac{dy}{dx} = \dfrac{4x}{\sqrt{x^2 - 5}}$ so $dy = \dfrac{4x}{\sqrt{x^2 - 5}}\, dx$

$\displaystyle\int dy = \int \dfrac{4x}{\sqrt{x^2 - 5}}\, dx$

Let $u = x^2 - 5$, so $du = 2x\,dx$ and $2\,du = 4x\,dx$

$\displaystyle\int \dfrac{4x}{\sqrt{x^2 - 5}}\, dx = \int u^{(-1/2)} \cdot 2\, du$

$\qquad\qquad\qquad = 4u^{(1/2)} + C$

$\qquad\qquad\qquad = 4\sqrt{x^2 - 5} + C$

$y = 4\sqrt{x^2 - 5} + C$

Using $y(3) = 6$, $6 = 4\sqrt{3^2 - 5} + C$

$6 = 4(2) + C \Rightarrow C = -2$

$y = 4\sqrt{x^2 - 5} - 2$

3. $4\dfrac{1}{3}$

Integrating the velocity function will calculate displacement from 1 to 3 seconds, which will cause a change in position at 1 second, $s(1) = 5$.

$s(3) = s(1) + \displaystyle\int_1^3 v(t)\,dt$

$\qquad = 5 + \displaystyle\int_1^3 (4 - t^2)\,dt$

$\qquad = 5 + \left(4t - \dfrac{1}{3}t^3\right)\Big|_1^3$

$\qquad = 5 + \left(4(3) - \dfrac{1}{3} \cdot 27\right) - \left(4(1) - \dfrac{1}{3} \cdot 1\right)$

$\qquad = 5 + 12 - 9 - 4 + \dfrac{1}{3}$

$\qquad = 4\dfrac{1}{3}$

4. Positive

   The velocity at $t = 6$ will be positive because the area under the acceleration graph causes a change in velocity. If $v(0) = -1$, for the first 2 seconds velocity will decrease about $1\frac{1}{2}$ more units (the area under the t-axis) to $-2\frac{1}{2}$. From $t = 2$ to 6 the area under the acceleration graph and above the t-axis will add at least 5 units of positive velocity.

5. 5 seconds

   Acceleration due to gravity is $\dfrac{dv}{dt} = -32\,\dfrac{\text{ft}}{\text{s}^2}$. Also, initial velocity is $v(0) = 64$, and initial position is $s(0) = 80$.

   $\displaystyle\int dv = \int -32\,dt$

   $v = -32t + C_1$

   Using $v(0) = 64$, $64 = -32(0) + C_1 \Rightarrow C_1 = 64$

   $\dfrac{ds}{dt} = v = -32t + 64$, so $ds = (-32t + 64)dt$

   $\displaystyle\int ds = \int (-32t + 64)\,dt$

   $s = -16t^2 + 64t + C_2$.

   Using $s(0) = 80$, $80 = -16(0)^2 + 64(0) + C_2 \Rightarrow C_2 = 80$

   $s = -16t^2 + 64t + 80$

   The object will hit the ground when $s = 0$, so $0 = -16(t^2 - 4t - 5)$

   $0 = -16(t - 5)(t + 1) \Rightarrow t = 5$

6. $P = 2e^{[\ln(\sqrt[4]{32})\cdot t]} = 2\cdot(\sqrt[4]{32})^t$

   Recognizing the phrase, "A population of bacteria grow at a rate proportional to the size of the population," jump to the exponential model, $P = P_0 e^{(k\cdot t)}$

   Day 0 is the initial population, $P_0 = 2$, so $P = 2e^{(k\cdot t)}$.

Use $P(4) = 64$ to solve for k.

$64 = 2e^{(k \cdot 4)} \Rightarrow 32 = e^{(4k)}$

$\ln(32) = \ln \left[e^{(4k)} \right]$

$\ln(32) = 4k \Rightarrow k = \dfrac{1}{4} \ln(32) = \ln(\sqrt[4]{32})$

$P = 2e^{[\ln(\sqrt[4]{32}) \cdot t]} = 2 \cdot (\sqrt[4]{32})^t$

7. $a = 4$

$$\int_0^2 [(t+1)^2 - a]\,dt = \left[\frac{1}{3}(t+1)^3 - a \cdot t \right]\Big|_0^2$$

$$= \left[\frac{1}{3}(3)^3 - 2a \right] - \left[\frac{1}{3}(1)^3 - 0a \right]$$

$$= \frac{26}{3} - 2a$$

$\dfrac{2}{3} = \dfrac{26}{3} - 2a \Rightarrow \dfrac{-24}{3} = -2a$

$-8 = -2a \Rightarrow a = 4$

8. 1

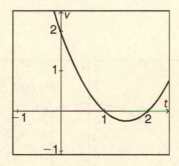

Figure 7.10

The graph of $v(t) = t^2 - 3t + 2$ is shown in Figure 7.10. Velocity is positive on $[0, 1]$, and negative on $[1, 2]$. To find total distance, the negative velocity will be made positive by reversing the limits of integration.

Total distance $= \int_0^1 (t^2 - 3t + 2)\, dt + \int_2^1 (t^3 - 3t + 2)\, dt$

$$= \left(\frac{1}{3} t^3 - \frac{3}{2} t^2 + 2t \right) \Big|_0^1 + \left(\frac{1}{3} t^3 - \frac{3}{2} t^2 + 2t \right) \Big|_2^1$$

$$= \left(\frac{1}{3} - \frac{3}{2} + 2 \right) + \left(\frac{1}{3} - \frac{3}{2} + 2 \right) - \left(\frac{1}{3} \cdot 2^3 - \frac{3}{2} \cdot 2^2 + 2(2) \right)$$

$$= \frac{2}{3} - 3 + 4 - \frac{8}{3} + 6 - 4$$

$$= 1$$

9. $\dfrac{3\pi}{2} - \dfrac{2}{3\pi}$

$$\frac{1}{\frac{3\pi}{2} - 0} \int_0^{\left(\frac{3\pi}{2}\right)} [2x + \cos(x)\, dx] = \frac{2}{3\pi} \cdot [x^2 + \sin(x)]\Big|_0^{(3\pi/2)}$$

$$= \frac{2}{3\pi} \cdot \left[\left(\frac{3\pi}{2} \right)^2 + \sin\left(\frac{3\pi}{2} \right) - (0^2 + \sin(0)) \right]$$

$$= \frac{2}{3\pi} \cdot \left[\frac{9\pi^2}{4} - 1 \right]$$

$$= \frac{3\pi}{2} - \frac{2}{3\pi}$$

10. 3π

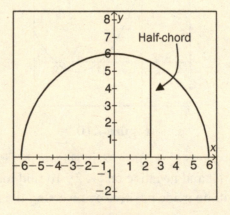

Figure 7.11

This problem is best done geometrically. Find the average value of a semicircle of radius 6 (see Figure 7.11), and double the result because the function values on the semicircle are half the chord lengths. Average value is the area of the semicircle divided by the width of the interval.

$$\frac{1}{6-(-6)} \cdot \left(\frac{1}{2}\pi \cdot 6^2\right) = \frac{1}{12} \cdot 18\pi$$

$$= \frac{3\pi}{2}$$

Doubling the average value gives an average chord length in a full circle of 3π.

11. No.

The average value is the net area divided by the width of the interval. If a small subset of the domain, less than half, contains a very large amount of area under the x-axis, and the rest of the domain has a very small amount of positive area, the average value can be negative.

12. $20 - 2\ln(5)$

On the interval $[0, 4]$, $f(x) = \dfrac{1}{2}x + 4$ is always above $h(x) = \dfrac{2}{x+1}$.

Area = $\displaystyle\int_0^4 [f(x) - h(x)]\,dx$.

$$\int_0^4 [f(x)-h(x)]\,dx = \int_0^4 \left[\frac{1}{2}x + 4 - \frac{2}{x+1}\right]dx$$

$$= \left[\frac{1}{4}x^2 + 4x - 2\ln(x+1)\right]\Big|_0^4$$

$$= \left[\frac{1}{4}\cdot 4^2 + 4(4) - 2\ln(4+1)\right] - \left[\frac{1}{4}\cdot 0^2 + 4(0) - 2\ln(0+1)\right]$$

$$= 20 - 2\ln(5)$$

13. 32

$g(x) = \dfrac{1}{4}x^3 - x$ and $h(x) = 3x$ are odd functions with symmetry to the origin.

$g(x) = \dfrac{1}{4}x^3 - x = 0$ so x-intercepts are $x = -2, 0, 2$.

A sketch of the graphs is shown in Figure 7.12.

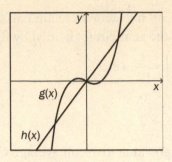

Figure 7.12

Find the points of intersection.

$$\frac{1}{4}x^3 - x = 3x$$

$$\frac{1}{4}x^3 = 4x$$

$$x^3 = 16x$$

$$x^3 - 16x = 0 \Rightarrow x(x-4)(x+4) = 0$$

Total Area $= 2\int_0^4 [h(x) - g(x)]\,dx$

$$2\int_0^4 \left[3x - \left(\frac{1}{4}x^3 - x\right)\right] dx = 2\int_0^4 \left(4x - \frac{1}{4}x^3\right) dx$$

$$= 2 \cdot \left(2x^2 - \frac{1}{16}x^4\right)\Big|_0^4$$

$$= 2 \cdot \left(2 \cdot 4^2 - \frac{1}{16} \cdot 4^4\right)$$

$$= 2(32 - 16)$$

$$= 32$$

14. $\sqrt{2} - 1$

The region is drawn in Figure 7.13.

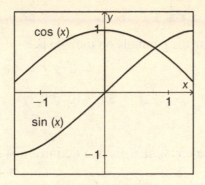

Figure 7.13

The desired point of intersection of $\sin(x)$ and $\cos(x)$ is $x = \dfrac{\pi}{4}$.

$$\int_0^{\frac{\pi}{4}} [\cos(x) - \sin(x)\, dx = [\sin(x) + \cos(x)] \Big|_0^{(\pi/4)}$$

$$= \left[\sin\left(\frac{\pi}{4}\right) + \cos\left(\frac{\pi}{4}\right) \right] - [\sin(0) + \cos(0)]$$

$$= \left[\frac{\sqrt{2}}{2} + \frac{\sqrt{2}}{2} \right] - [0 + 1]$$

$$= \sqrt{2} - 1$$

15. $\displaystyle\int_{-2}^{1} [x + 2 - (x^2 - 4)]\, dx + \int_{1}^{2} [6 - 3x - (x^2 - 4)]\, dx$

A sketch of the graphs is shown in Figure 7.14. Each graph should be drawn by hand.

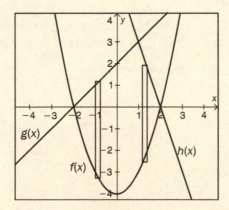

Figure 7.14

$x + 2 = 6 - 3x \Rightarrow x = 1$

Each line intersects the parabola on the x-axis.

$$x^2 - 4 = x + 2 \qquad\qquad x^2 - 4 = 6 - 3x$$

$$x^2 - x - 6 = 0 \qquad\qquad x^2 + 3x - 10 = 0$$

$$x = -2 \qquad\qquad x = 2$$

The heights of the rectangles change definition at $x = 1$.

$$\int_{-2}^{1} [g(x) - f(x)] \, dx + \int_{1}^{2} [h(x) - f(x)] \, dx =$$

$$\int_{-2}^{1} [(x+2) - (x^2 - 4)] \, dx + \int_{1}^{2} [(6 - 3x) - (x^2 - 4)] \, dx$$

PRACTICE TEST 1

CLEP Calculus

Also available at the REA Study Center (*www.rea.com/studycenter*)

This practice test is also offered online at the REA Study Center. All CLEP exams are administered on computer, and our test is formatted to simulate test-day conditions. We recommend that you take the online version of this test to receive these added benefits:

- **Timed testing conditions** – helps you gauge how much time you can spend on each question
- **Automatic scoring** – find out how you did on the test, instantly
- **On-screen detailed explanations of answers** – gives you the correct answer and explains why the other answer choices are wrong
- **Diagnostic score reports** – pinpoint where you're strongest and where you need to focus your study

PRACTICE TEST 1

CLEP Calculus

(Answer sheets appear in the back of the book.)

TOTAL TIME: 90 Minutes
44 Questions

Section I

TIME: 50 Minutes
27 Questions

Directions: Solve each of the following problems without using a calculator. Choose the best answer from those provided. Some questions will require you to enter a numerical answer in the box provided.

Notes: (1) Figures that accompany questions are intended to provide information useful in answering the questions. All figures lie in a plane unless otherwise indicated. The figures are drawn as accurately as possible EXCEPT when it is stated in a specific question that the figure is not drawn to scale. Straight lines and smooth curves may appear slightly jagged.

(2) Unless otherwise specified, all angles are measured in radians, and all numbers used are real numbers.

(3) Unless otherwise specified, the domain of any function f is assumed to be the set of all real numbers x for which $f(x)$ is a real number. The range of f is assumed to be the set of all real numbers $f(x)$, where x is the domain of f.

(4) In this exam, $\ln x$ denotes the natural logarithm of x (the logarithm to the base e).

(5) The inverse of a trigonometric function f may be indicated using the inverse function notation f^{-1} or with the prefix "arc" (e.g., $\sin^{-1} x = \arcsin x$).

1. If $y = 6x^2 + x$, then $\dfrac{dy}{dx} =$

   (A) $12x + 1$

   (B) $12x$

   (C) $2x^3 + \dfrac{1}{2}x^2$

   (D) $13x$

   (E) $8x + 1$

2. $\displaystyle\int \cos(3x)\,dx =$

   (A) $\dfrac{x \cdot \sin(3x)}{3} + C$

   (B) $-3\sin(3x) + C$

   (C) $3\sin(3x) + C$

   (D) $-\dfrac{\sin(3x)}{3} + C$

   (E) $\dfrac{\sin(3x)}{3} + C$

3. The velocity of a bicycle is constantly increasing, as shown in the table below.

Time (sec.)	1	3	6	7	9
Velocity (ft./sec.)	5	7	10	14	20

   Using a Riemann sum with four subdivisions, what is the upper estimate of the number of feet traveled?

4. What is $\lim\limits_{x\to\infty} \dfrac{3x^2 - x}{4x + 5x^2}$?

(A) 0

(B) $\dfrac{2}{9}$

(C) $\dfrac{3}{5}$

(D) $\dfrac{3}{4}$

(E) The limit does not exist.

5. If $w(x) = x^{\left(\frac{3}{2}\right)}$, what is the instantaneous rate of change of w with respect to x at $x = 16$?

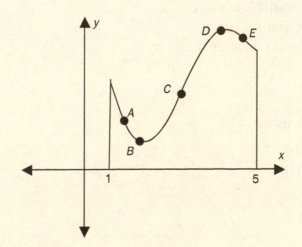

6. The function h is shown in the figure above. At which of the following points is h' equal to the average rate of change of h over the interval [1, 5]?

(A) A

(B) B

(C) C

(D) D

(E) E

7. Function f is continuous over all real numbers, and a, b, and c are positive real constants. $\int_a^b f(2x-c)\,dx$ is equivalent to

   (A) $\dfrac{1}{2}\displaystyle\int_{2a-c}^{2b-c} f(u)\,du$

   (B) $2\displaystyle\int_{2a-c}^{2b-c} f(u)\,du$

   (C) $\displaystyle\int_{2a-c}^{2b-c} f(u)\,du$

   (D) $\dfrac{1}{2}\displaystyle\int_{2a+c}^{2b+c} f(u)\,du$

   (E) $2\displaystyle\int_{2a+c}^{2b+c} f(u)\,du$

8. Let g be a function defined over all real numbers and a be a constant. If $\lim\limits_{x \to a} g(x) = g(a)$ which of the following statements MUST be true?

   I. g is differentiable at $x = a$
   II. g is continuous at $x = a$
   III. g has a local minimum at $x = a$

   (A) I only
   (B) II only
   (C) I and II only
   (D) II and III only
   (E) I, II, and III

9. $\dfrac{d}{dx}\displaystyle\int_2^x e^{(t^3)}\,dt =$

   (A) $e^{(x^3)}$
   (B) $e^{(x^3)} - e^8$
   (C) $\dfrac{e^{(x^3)}}{3x^2}$
   (D) $3x^2 e^{(x^3)}$
   (E) $e^{(3x^2)} - e^{12}$

10. If $y = \dfrac{x^2}{x+1}$, then $\dfrac{dy}{dx} =$

   (A) $2x$

   (B) $\dfrac{2x}{x+1}$

   (C) $\dfrac{x^2 + 2x}{(x+1)^2}$

   (D) $\dfrac{-x^2 - 2x}{(x+1)^2}$

   (E) $\dfrac{3x^2 + 2x}{(x+1)^2}$

11. For $h(x) = \sqrt{x^3}$, find the value of x where $\dfrac{dh}{dx} = \dfrac{d^2h}{dx^2}$.

   (A) 0

   (B) $\dfrac{1}{4}$

   (C) $\dfrac{1}{2}$

   (D) $\dfrac{\sqrt{3}}{2}$

   (E) 1

12. The Riemann sum $\dfrac{1}{30} \displaystyle\sum_{k=1}^{60} e^{\left(\frac{k}{30}\right)}$ is an approximation for which of the following integrals?

   (A) $\displaystyle\int_0^2 e^{\left(\frac{x}{30}\right)} dx$

   (B) $\displaystyle\int_0^2 e^x \, dx$

   (C) $\displaystyle\int_0^{60} e^x \, dx$

   (D) $\dfrac{1}{30} \displaystyle\int_0^2 e^x \, dx$

   (E) $\dfrac{1}{30} \displaystyle\int_0^{60} e^{\left(\frac{x}{30}\right)} dx$

13. What is $\lim\limits_{x \to 5} \dfrac{x^2 - 2x - 15}{x - 5}$?

   (A) 0
   (B) 1
   (C) 2
   (D) 8
   (E) The limit does not exist.

14. If $y = x^3 \cdot \ln(x)$, then $y' =$

   (A) $3x$
   (B) $3x^2 + \dfrac{1}{x}$
   (C) $3x^3$
   (D) $3x^2 \ln(x) + x^3$
   (E) $x^2[3 \cdot \ln(x) + 1]$

x	$f(x)$	$f'(x)$	$g(x)$	$g'(x)$
1	4	-3	0	-6
2	2	-1	-2	7
3	3	5	1	2

15. Given $h(x) = f(g(x))$, use the table of values above to find $h'(3)$.

   (A) -6
   (B) -3
   (C) -1
   (D) 0
   (E) 10

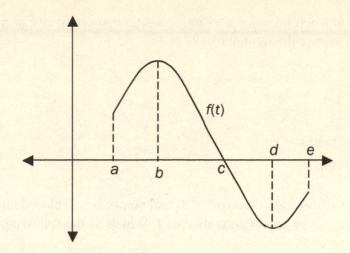

16. The graph of $f(t)$ is given above. a, b, c, d and e are real constants. If $g(x) = \int_a^x f(t)\,dt$, on which interval is the graph of $g(x)$ concave down?

    (A) (a, c)
    (B) (a, b)
    (C) (c, e)
    (D) $(a, b) \cup (d, e)$
    (E) (b, d)

17. Let functions f and g be inverses. If $y = 3x - 1$ is tangent to g at $x = 2$, which of the following statements MUST be true?

    I. f is decreasing at $x = 2$

    II. $f'(5) = \dfrac{1}{3}$

    III. $f'(2) = 3$

    (A) I only
    (B) II only
    (C) III only
    (D) I & II only
    (E) II & III only

18. Which function provides a visual example that continuity at a point does NOT guarantee differentiability?

   (A) $y = x^2$

   (B) $y = |x|$

   (C) $y = 2^x$

   (D) $y = \tan(x)$

   (E) $y = \ln(x)$

19. Functions f and h are defined for all real numbers on a closed interval $[a, b]$, with $a < b$. h is an antiderivative of f. Which of the following statements MUST be true?

   (A) If $h < 0$ on the entire interval, then $f < 0$ on the entire interval.

   (B) $h'(x) = f(x)$ for all real numbers, x in $[a, b]$.

   (C) If f is constant, then $h(x) = 0$ for all real numbers, x in $[a, b]$.

   (D) If $f(c) = 0$ for some c in (a, b), then h has a local extrema at $x = c$.

   (E) If $f > 0$ on the interval $[a, b]$, then $h(c) = 0$ for some c in (a, b).

20. $\int \tan^2 x \, dx =$

   (A) $\dfrac{\tan^3 x}{3} + C$

   (B) $\dfrac{\tan^3 x}{3 \sec^2 x} + C$

   (C) $x - \tan(x) + C$

   (D) $\tan(x) - x + C$

   (E) $\dfrac{\sec^3 x}{3} - x + C$

21. On the interval $[a, c]$, $\int_a^b f(x)\,dx = -2$ and $\int_a^c f(x)\,dx = 8$. Given that $a < b < c$, evaluate $\int_b^c 2f(x)\,dx$.

   (A) 20

   (B) 16

   (C) 12

   (D) 10

   (E) 8

22. If $h(x) = \dfrac{\sin(x)}{\cos(x)}$, then $h'(x) =$

    (A) $1 - \tan^2(x)$

    (B) $-\dfrac{\cos(x)}{\sin(x)}$

    (C) $\dfrac{\cos(x)}{\sin(x)}$

    (D) $\sec^2(x)$

    (E) 1

23. What is $\displaystyle\lim_{h \to 0} \dfrac{(2+h)^5 - 2^5}{h}$?

    (A) 0

    (B) 1

    (C) 32

    (D) 80

    (E) The limit does not exist.

24. If $y = \arctan(x^2)$, $\dfrac{dy}{dx} =$

    (A) $\dfrac{1}{1+x^4}$

    (B) $\dfrac{2x}{1+x^4}$

    (C) $\dfrac{1}{1+x^2}$

    (D) $\dfrac{2x}{1+x^2}$

    (E) $\dfrac{2x}{x^2+x^4}$

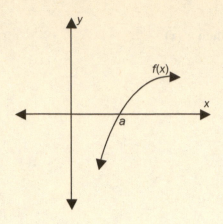

25. The figure above shows the graph of a twice differentiable function, f, crossing the x-axis at $x = a$. Which compound inequality is true?

    (A) $f(a) < f'(a) < f''(a)$
    (B) $f(a) < f''(a) < f'(a)$
    (C) $f'(a) < f(a) < f''(a)$
    (D) $f''(a) < f'(a) < f(a)$
    (E) $f''(a) < f(a) < f'(a)$

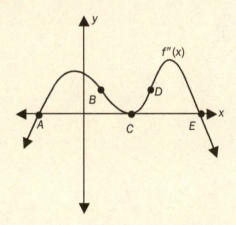

26. The graph of $f''(x)$ is shown above. The graph is tangent to the x-axis at point C. Where does the graph of f have an inflection point?

    (A) A & E
    (B) B & D
    (C) A, C & E
    (D) A only
    (E) E only

27. Which graph below could be the graph of a solution to $\dfrac{dy}{dx} = k \cdot y$ if k is a negative constant?

(A)

(B)

(C)

(D)

(E)

Section II

TIME: 40 Minutes
   17 Questions

Directions: A graphing calculator is available for the questions in this section. Choose the best answer from those provided. Some questions will require you to enter a numerical answer in the box provided.

28. Find the value of b such that $\int_{1}^{b} 2x^{-1} \, dx = 5$.

    (A) e^3

    (B) $e^{2.5}$

    (C) e^2

    (D) 6

    (E) 3

29. The acceleration of an object is always 4 feet per second squared. If at time $t = 1$ the object is stationary and has position, $x(1) = 5$ feet, find the position function, $x(t)$.

    (A) $x(t) = 2t^2 - 2t + 5$

    (B) $x(t) = 2t^2 + 4t - 1$

    (C) $x(t) = 2t^2 + 3$

    (D) $x(t) = 2t^2 - 4t + 7$

    (E) $x(t) = 2t^2 - 4t + 5$

30. Write the equation of the line tangent to $x^2y + y^3 + 4x = 8$ at the point where the given curve intersects the x-axis.

    (A) $y = x - 2$

    (B) $y = -x + 2$

    (C) $y = 0$

    (D) $y = \dfrac{-1}{3}x + 2$

    (E) $y = \dfrac{1}{3}x + 2$

31. Let $f(x) = \frac{1}{3} x^3 + x + 2$. On the interval $[0, 3]$, find the value(s) of x where the average rate of change of f equals the instantaneous rate of change of f.

    (A) $\sqrt{5}$

    (B) ± 1

    (C) 1

    (D) $\pm\sqrt{3}$

    (E) $\sqrt{3}$

32. Use four trapezoids to estimate the area under the graph of $f(x) = 1 + 2 \sin(x)$ and above the x-axis on the interval $[0, \pi]$.

    (A) $\pi(3 + \sqrt{2})$

    (B) $\frac{\pi}{4}(7 + 2\sqrt{2})$

    (C) $\pi(1 + \sqrt{2})$

    (D) $\frac{\pi}{2}(1 + \sqrt{2})$

    (E) $\frac{\pi}{2}(3 + \sqrt{2})$

33. Find the area enclosed by the graphs of $y = 4 - x$ and $y = (x - 2)^2$.

    (A) $\frac{9}{2}$

    (B) $\frac{31}{6}$

    (C) $\frac{59}{6}$

    (D) $\frac{21}{2}$

    (E) $\frac{27}{2}$

34. The absolute minimum value of $g(x) = 2x^3 - \dfrac{7}{2} x^2 - 3x + 1$ on the interval $[-1, 2]$ occurs at $x =$

    (A) -1

    (B) $\dfrac{-1}{3}$

    (C) 1

    (D) $\dfrac{3}{2}$

    (E) 2

35. The acceleration of a particle as a function of time is given by $a(t) = 2^t \cdot \ln 4$. At time $t = 0$, the velocity of the particle is 5 feet per second. What is the velocity of the particle, in feet per second, at time $t = 3$ seconds?

    (A) $5 + 2\ln(4)$

    (B) 13

    (C) 14

    (D) $5 + \dfrac{7}{\ln(2)}$

    (E) 19

36. Find the average value of $f(x) = x \cdot e^{(x^2)}$ on the interval $[0, 2]$.

    (A) $\dfrac{e^4 - 1}{4}$

    (B) $\dfrac{e^4}{4}$

    (C) $\dfrac{e^4 - 1}{2}$

    (D) $e^4 - 1$

    (E) e^4

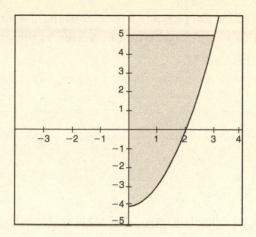

37. Find the total area of the shaded region on the graph above. The function is $h(x) = x^2 - 4$ on $[\, 0, 3 \,]$.

    (A) $15\dfrac{1}{3}$

    (B) $17\dfrac{2}{3}$

    (C) 18

    (D) $20\dfrac{1}{3}$

    (E) $24\dfrac{2}{3}$

$f(2)$	$f'(2)$	$g(2)$	$g'(2)$
3	5	8	−4

38. If $h(x) = f(x) \cdot g(x)$, use the table above to determine the value of $h'(2)$.

39. Find the area of the largest possible isosceles triangle which can be inscribed under the graph of $y = 9 - x^2$ with one vertex at the origin and its upper base parallel to the x-axis.

    (A) $\sqrt{3}$

    (B) $3\sqrt{3}$

    (C) 9

    (D) $6\sqrt{3}$

    (E) 12

40. The velocity of an object is given by $v(t) = t \cdot (t^2 - 1)^{\left(\frac{1}{3}\right)}$ feet per second. Find the total number of feet traveled by the object in the time interval $[0, 3]$ seconds.

    (A) $\dfrac{51}{4}$

    (B) $\dfrac{34}{3}$

    (C) $\dfrac{45}{4}$

    (D) $\dfrac{51}{8}$

    (E) $\dfrac{45}{8}$

41. Find the slope of the line tangent to $f(x) = \sin(x^3 + x)$ at $x = 0$.

    (A) 0

    (B) 1

    (C) 2

    (D) 3

    (E) 4

42. Let h be differentiable over all real numbers with $h'(x) = x(x-2)^2(x+1)$. On which interval(s) is h decreasing?

   (A) $[-1, 0]$
   (B) $(-\infty, -1]$ and $[0, \infty)$
   (C) $(-\infty, -1]$ and $[0, 2]$
   (D) $[0, 2]$
   (E) All real numbers

43. Let A be the number of ants in a colony as a function of time t months. If $\dfrac{dA}{dt} = 0.4A$ and $A(0) = 12000$, find $A(t)$.

   (A) $\ln(12000)e^{0.4t}$
   (B) $12000 + e^{0.4t}$
   (C) $12000\, e^{0.4t}$
   (D) $12000 - 0.4t$
   (E) $12000 + 0.4t^2$

44. The position of a particle moving along the x-axis as a function of time, t, is given by $x(t) = 12t - t^3$ for $t \geqslant 0$. What is the particle's acceleration at the instant the particle is farthest to the right?

   (A) -12
   (B) -6
   (C) 0
   (D) 2
   (E) 16

PRACTICE TEST 1

Answer Key

1. (A)	16. (E)	31. (E)
2. (E)	17. (B)	32. (E)
3. 98	18. (B)	33. (A)
4. (C)	19. (B)	34. (D)
5. 6	20. (D)	35. (E)
6. (D)	21. (A)	36. (A)
7. (A)	22. (D)	37. (C)
8. (B)	23. (D)	38. 28
9. (A)	24. (B)	39. (D)
10. (C)	25. (E)	40. (D)
11. (C)	26. (A)	41. (B)
12. (B)	27. (B)	42. (A)
13. (D)	28. (B)	43. (C)
14. (E)	29. (D)	44. (A)
15. (A)	30. (B)	

PRACTICE TEST 1

Detailed Explanations of Answers

1. **(A)**   Utilize the derivative formula, $\dfrac{d}{dx}(x^n) = n \cdot x^{n-1}$

$$\frac{d}{dx}(6x^2 + x) = (6)(2)x^{2-1} + 1x^{1-1}$$
$$= 12x + 1$$

2. **(E)**   Let $u = 3x$ and $du = 3dx$ which means $dx = \dfrac{1}{3}du$. Substituting into the integrand,

$$\int \cos(3x)\,dx = \frac{1}{3}\int \cos(u)\,du$$
$$\frac{1}{3}\int \cos(u)\,du = \frac{1}{3}\sin(u) + C$$
$$= \frac{1}{3}\sin(3x) + C$$

3. **98**   Since the velocity is constantly increasing, the right-oriented rectangle will always use the greatest velocity in each interval guaranteeing an upper estimate of distance traveled. Notice time intervals are not equal.

Distance $\approx 7(3-1) + 10(6-3) + 14(7-6) + 20(9-7)$

$14 + 30 + 14 + 40 = 98$

4. **(C)**

$$\lim_{x \to \infty} \frac{3x^2 - x}{4x + 5x^2} = \lim_{x \to \infty} \frac{3x^2 - x}{4x + 5x^2} \cdot \frac{\frac{1}{x^2}}{\frac{1}{x^2}}$$
$$= \lim_{x \to \infty} \frac{3 - \frac{1}{x}}{\frac{4}{x} + 5}$$
$$= \frac{3 - 0}{0 + 5}$$
$$= \frac{3}{5}$$

5. **6** The instantaneous rate of change when $x = 16$ is the value of the derivative.

$$\frac{d}{dx} x^{\left(\frac{3}{2}\right)}\bigg|_{x=16} = \frac{3}{2} x^{\left(\frac{1}{2}\right)}\bigg|_{x=16}$$

$$= \frac{3}{2} \cdot 16^{\left(\frac{1}{2}\right)}$$

$$= 6$$

6. **(D)** The average rate of change is the slope of a segment connecting the endpoints of the curve on the interval $[1, 5]$. h' is the slope of a tangent to the curve. Point D is the only given point where the slope of the tangent is equal to the slope of the segment.

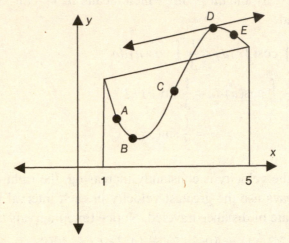

7. **(A)** Let $u = 2x - c$, $du = 2dx$, or $dx = \frac{1}{2} du$.

When $x = a$, $u = 2a - c$ and when $x = b$, $u = 2b - c$

By substitution, $\int_{a}^{b} f(2x - c)\, dx = \frac{1}{2} \int_{2a-c}^{2b-c} f(u)\, du$

8. **(B)** $\lim\limits_{x \to a} g(x) = g(a)$ is the necessary condition for continuity at a point, a. We know statement II is true. Differentiability at a point guarantees continuity, but continuity does not guarantee differentiability. The function g could have a cusp at $(a, g(a))$. Statement III does not have to be true since a cusp at a high point would be a local maximum, not a minimum. Only statement II must be true.

9. **(A)** Use the Fundamental Theorem of Calculus, $\dfrac{d}{dx}\displaystyle\int_a^x f(t)\,dt = f(x)$.

$$\frac{d}{dx}\int_2^x e^{(t^3)}\,dt = e^{(x^3)}$$

10. **(C)** Use the quotient rule, $d\left(\dfrac{u}{v}\right) = \dfrac{v\cdot u' - u\cdot v'}{v^2}$

$$\frac{d}{dx}\left(\frac{x^2}{x+1}\right) = \frac{(x+1)\cdot 2x - x^2(1)}{(x+1)^2}$$

$$= \frac{2x^2 + 2x - x^2}{(x+1)^2}$$

$$= \frac{x^2 + 2x}{(x+1)^2}$$

11. **(C)** For $h(x) = \sqrt{x^3}$, find $\dfrac{dh}{dx}$ and $\dfrac{d^2h}{dx^2}$, set them equal and solve. It is much easier to recognize $h(x) = x^{\left(\frac{3}{2}\right)}$.

$$\frac{dh}{dx} = \frac{3}{2}x^{\left(\frac{1}{2}\right)}$$

$$\frac{d^2h}{dx^2} = \frac{3}{4}x^{\left(\frac{-1}{2}\right)}$$

$$\frac{3\sqrt{x}}{2} = \frac{3}{4\sqrt{x}}$$

$$12x = 6$$

$$x = \frac{1}{2}$$

12. **(B)** Identify $\dfrac{1}{30}$ as Δx and each term of the expanded summation notation as $f(x_k)$.

The expansion becomes,

$$e^{\left(\frac{1}{30}\right)}\cdot\frac{1}{30} + e^{\left(\frac{2}{30}\right)}\cdot\frac{1}{30} + e^{\left(\frac{3}{30}\right)}\cdot\frac{1}{30} + \ldots + e^{\left(\frac{60}{30}\right)}\cdot\frac{1}{30}.$$

This is a right oriented rectangular approximation for e^x on the interval $[0, 2]$ with the first used function value $e^{\left(\frac{1}{30}\right)}$ and the last used function value $e^{\left(\frac{60}{30}\right)} = e^2$. Using 60 rectangles would make the width of each rectangle, $\Delta x = \dfrac{2}{60} = \dfrac{1}{30}$.

$$\frac{1}{30}\sum_{k=1}^{60} e^{\left(\frac{k}{30}\right)} \approx \int_0^2 e^x \, dx.$$

13. **(D)**

$$\lim_{x \to 5} \frac{x^2 - 2x - 15}{x - 5} = \lim_{x \to 5} \frac{(x-5)(x+3)}{x-5}$$
$$\lim_{x \to 5}(x + 3) = 8.$$

14. **(E)** Use product rule to find $\dfrac{dy}{dx}$ for $y = x^3 \cdot \ln(x)$.

$$\frac{dy}{dx} = \frac{d}{dx}[x^3 \cdot \ln(x)]$$
$$= \frac{d}{dx}(x^3) \cdot \ln(x) + \frac{d}{dx}\ln(x) \cdot x^3$$
$$= 3x^2 \cdot \ln(x) + \frac{1}{x} \cdot x^3$$
$$= 3x^2 \cdot \ln(x) + x^2$$
$$= x^2[3\ln(x) + 1]$$

15. **(A)** Using the chain rule on $h(x) = f(g(x))$, $h'(x) = f'(g(x)) \cdot g'(x)$
$$h'(3) = f'(g(3)) \cdot g'(3)$$
$$= f'(1) \cdot g'(3)$$
$$= -3 \cdot 2 = -6$$

16. **(E)** Using the Fundamental Theorem of Calculus, if $g(x) = \displaystyle\int_a^x f(t)\,dt$, then $g'(x) = f(x)$ and $g''(x) = f'(x)$. g is concave down when $g''(x) < 0$ or when $f'(x) < 0$. $f'(x) < 0$ where f is decreasing, which is on the interval (b, d).

17. **(B)** Use the rule for slopes of inverse functions, $f'(x) = \dfrac{1}{g'(f(x))}$.

If $y = 3x - 1$ is tangent to g at $x = 2$, $g'(2) = 3$, the slope of the line, and $g(2) = 3 \cdot 2 - 1 = 5$, the point common to the line and g.

Since f is the inverse of g, and $g(2) = 5$, then $f(5) = 2$, and the slope of the tangent to f at $(5, 2)$ is the reciprocal of $g'(2)$, so $f'(5) = \dfrac{1}{g'(2)} = \dfrac{1}{3}$.

Nothing can be determined about what is happening on the graph of f when $x = 2$.

18. **(B)** The graph of $y = |x|$ is continuous at $x = 0$, but it is not differentiable since there is a corner at $(0, 0)$.

19. **(B)** If h is an antiderivative of f, another way to write that is

$$\int f(x)\,dx = h(x) + C.$$

Using the Fundamental Theorem of Calculus, $f(x) = h'(x)$.

To rule out choice A, consider any negative function h which is always increasing.

Choice C is wrong since the antiderivative of a constant function is a linear function, and there is no guarantee it equals 0.

Since $f(x) = h'(x)$, there must be a change in sign of $h'(x)$ to guarantee an extrema, thus D is not necessarily true.

Thinking of h as the accumulated area under f, for all x in (a, b), $\int_a^x f(x)\,dx$ would be positive if $f > 0$. Therefore, E is not correct.

20. **(D)** Use a substitution from the trigonometric identity, $1 + \tan^2(x) = \sec^2 x$.

$$\int \tan^2 x\,dx = \int (\sec^2 x - 1)\,dx$$
$$\int (\sec^2 x - 1)\,dx = \int \sec^2 x\,dx - \int 1\,dx$$
$$= \tan(x) - x + C$$

21. **(A)**

$$\int_a^b f(x)\,dx + \int_b^c f(x)\,dx = \int_a^c f(x)\,dx$$

$$-2 + \int_b^c f(x)d(x) = 8$$

$$\int_b^c f(x)\,dx = 10$$

$$\int_b^c 2f(x)\,dx = 2\int_b^c f(x)\,dx$$

$$= 20$$

22. **(D)**

$$h(x) = \frac{\sin(x)}{\cos(x)} = \tan(x)$$

$$h'(x) = \frac{d}{dx}\tan(x) = \sec^2(x)$$

23. **(D)** By comparison to the definition of the derivative, $f'(x) = \lim\limits_{h \to 0} \dfrac{f(x+h) - f(x)}{h}$, it can be seen that x has been replaced by 2, and the function is $f(x) = x^5$. $\lim\limits_{h \to 0} \dfrac{(2+h)^5 - 2^5}{h}$ is the derivative of x^5 evaluated when $x = 2$.

$$\frac{d}{dx}(x^5) = 5x^4$$

$$5 \cdot 2^4 = 80$$

24. **(B)** If u is a function of x, $\dfrac{d}{dx}(\arctan(u)) = \dfrac{\frac{du}{dx}}{1 + u^2}$.

$$\frac{d}{dx}(\arctan(x^2)) = \frac{\frac{du}{dx}(x^2)}{1 + (x^2)^2} = \frac{2x}{1 + x^4}$$

25. **(E)** $f(a) = 0$

The graph of f is increasing at a, so $f'(a) > 0$.

The graph of f is concave down at a, so $f''(a) < 0$.

$$f''(a) < f(a) < f'(a)$$

26. **(A)** Notice that the given graph is f'' and not f. An inflection point on f occurs when f'' changes sign. This happens at points A and E on the given graph of f''.

27. **(B)** The familiar differential equation $\dfrac{dy}{dx} = k \cdot y$ integrates to $y = Ae^{kx}$ where A is a positive or negative constant determined by an initial condition. Since k is negative, the only graph which fits these conditions is graph B, an exponential decay graph.

28. **(B)** Find the value of b such that $\displaystyle\int_1^b 2x^{-1}\, dx = 5$.

$$\int_1^b 2x^{-1}\, dx = \int_1^b \frac{2}{x}\, dx$$
$$= 2\, \ln(x)\Big|_1^b$$
$$= 2[\ln(b) - \ln(1)]$$

$$2[\ln(b) - \ln(1)] = 5$$
$$2[\ln(b) - 0] = 5$$
$$\ln(b) = 2.5$$
$$b = e^{2.5}$$

29. **(D)** $a(t) = 4$, $v(1) = 0$, $x(1) = 5$

$v(t) = \displaystyle\int 4\, dt \Rightarrow v(t) = 4t + C$

Using $v(1) = 0$, $0 = (4)(1) + C$, so $C = -4 \Rightarrow v(t) = 4t - 4$

$x(t) = \displaystyle\int 4t - 4\, dt \Rightarrow x(t) = 2t^2 - 4t + K$

Using $x(1) = 5$, $5 = (2)(1)^2 - (4)(1) + K$ so $K = 7 \Rightarrow x(t) = 2t^2 - 4t + 7$

30. **(B)**

I. Substitute $y = 0$ into $x^2y + y^3 + 4x = 8$ to find the x-intercept.
$$x^2 \cdot 0 + 0^3 + 4x = 8 \text{ gives } x = 2.$$
The point of tangency is $(2, 0)$.

II. Find the slope, $\dfrac{dy}{dx}$, by implicit differentiation.

$$\frac{d}{dx}(x^2y + y^3 + 4x) = \frac{d}{dx}(8)$$
$$x^2 \cdot \frac{dy}{dx} + 2x \cdot y + 3y^2\frac{dy}{dx} + 4 = 0$$

III. Evaluate at (2, 0).

$$2^2 \cdot \frac{dy}{dx} + 2 \cdot 2 \cdot 0 + 3 \cdot 0^2 \frac{dy}{dx} + 4 = 0$$

$$4\frac{dy}{dx} + 4 = 0$$

$$\frac{dy}{dx} = -1$$

IV. Write the equation,

$$y - 0 = -1(x - 2)$$

$$y = -x + 2$$

31. **(E)** For $f(x) = \frac{1}{3}x^3 + x + 2$ on [0, 3], the average rate of change is $\frac{f(3) - f(0)}{3 - 0}$.

The instantaneous rate of change is $f'(x) = x^2 + 1$.

Setting them equal, $x^2 + 1 = \frac{14 - 2}{3}$.

$$x^2 + 1 = 4$$

$$x^2 = 3$$

$$x = \pm\sqrt{3}$$

Only $x = \sqrt{3}$ lies in the interval [0, 3].

32. **(E)** Use four trapezoids to estimate the area under the graph of $f(x) = 1 + 2\sin(x)$ and above the x-axis on the interval $[0, \pi]$.

$$\text{Area} \approx \frac{\Delta x}{2}[f(x_0) + 2f(x_1) + 2f(x_2) + 2f(x_3) + f(x_4)]$$

$$\text{Area} \approx \frac{\frac{\pi}{4}}{2}[f(0) + 2f\left(\frac{\pi}{4}\right) + 2f\left(\frac{2\pi}{4}\right) + 2f\left(\frac{3\pi}{4}\right) + f(\pi)]$$

$$\text{Area} \approx \frac{\pi}{8}\left[1 + 2\left(1 + 2 \cdot \frac{\sqrt{2}}{2}\right) + 2(1 + 2 \cdot 1) + 2\left(1 + 2 \cdot \frac{\sqrt{2}}{2}\right) + 1\right]$$

$$\text{Area} \approx \frac{\pi}{8}[1 + (2 + 2\sqrt{2}) + 6 + (2 + 2\sqrt{2}) + 1] = \frac{\pi}{8}(12 + 4\sqrt{2})$$

$$\text{Area} \approx \frac{\pi}{8}(12 + 4\sqrt{2}) = \frac{\pi}{8} \cdot 4(3 + \sqrt{2})$$

$$= \frac{\pi}{2}(3 + \sqrt{2})$$

33. **(A)** Sketch the graphs and find their points of intersection, $(0, 4)$ and $(3, 1)$.

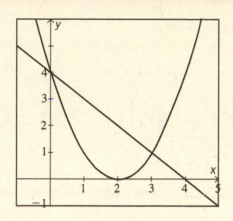

$$\text{Area} = \int_0^3 4 - x - (x-2)^2 \, dx$$

$$= 4x - \frac{x^2}{2} - \frac{(x-2)^3}{3} \bigg|_0^3$$

$$= \left(4 \cdot 3 - \frac{3^2}{2} - \frac{(3-2)^3}{3}\right) - \left(4 \cdot 0 - \frac{0^2}{2} - \frac{(0-2)^3}{3}\right)$$

$$= \left(12 - \frac{9}{2} - \frac{1}{3}\right) - \left(-\frac{-8}{3}\right)$$

$$= 12 - \frac{9}{2} - 3$$

$$= \frac{9}{2}$$

34. **(D)** Compare the local minimum to values at the endpoints of the interval $[-1, 2]$.

If $g(x) = 2x^3 - \frac{7}{2}x^2 - 3x + 1$, then $g'(x) = 6x^2 - 7x - 3$.

$$6x^2 - 7x - 3 = 0$$

$$(2x - 3)(3x + 1) = 0$$

$$x = \frac{3}{2} \text{ or } x = \frac{-1}{3}$$

$g''(x) = 12x - 7$ and $g''\left(\dfrac{3}{2}\right) = 11 > 0$ so a local minimum exists at $x = \dfrac{3}{2}$ since $g'(x) = 0$ and $g''(x) > 0$.

$$g(-1) = -2 - \dfrac{7}{2} + 3 + 1 = -1\dfrac{1}{2}$$

$$g\left(\dfrac{3}{2}\right) = 2\left(\dfrac{27}{8}\right) - \dfrac{7}{2}\left(\dfrac{9}{4}\right) - 3\left(\dfrac{3}{2}\right) + 1$$

$$= \dfrac{27}{4} - \dfrac{63}{8} - \dfrac{9}{2} + 1$$

$$= \dfrac{-37}{8}$$

$g(2) = 16 - 14 - 6 + 1 = -3$

Absolute minimum is at $x = \dfrac{3}{2}$.

Note that $g''\left(-\dfrac{1}{3}\right) = -11 < 0$, so that a local maximum exists at $x = -\dfrac{1}{3}$ since $g'(x) = 0$ and $g''(x) < 0$. In fact, $g\left(-\dfrac{1}{3}\right) = \dfrac{83}{54}$, which is the highest value of $g(x)$ in the interval $[-1, 2]$.

35. **(E)**   The definite integral of acceleration is the net change in velocity during a given time interval. Add the net change to the initial velocity.

$$v(3) = v(0) + \int_0^3 a(t)\, dt$$

$$= 5 + \int_0^3 2^t \ln(4)\, dt$$

$$= 5 + \ln(4)\dfrac{2^t}{\ln(2)}\Bigg|_0^3$$

$$= 5 + \dfrac{2\ln(2)}{\ln(2)}(2^3 - 2^0)$$

$$= 5 + 2(8 - 1)$$

$$= 19$$

36. **(A)** Find the average value of $f(x) = x \cdot e^{(x^2)}$ on the interval $[0, 2]$.

$$f_{avg} = \frac{1}{b-a} \int_a^b f(x)\,dx$$

$$f_{avg} = \frac{1}{2-0} \int_0^2 x \cdot e^{(x^2)}\,dx$$

Let $u = x^2$ and $du = 2x\,dx$ or $x\,dx = \frac{1}{2}\,du$.

If $x = 0$, $u = 0$, and if $x = 2$, then $u = 4$.

$$f_{avg} = \frac{1}{2-0} \int_0^4 e^{(u)} \cdot \frac{1}{2}\,du$$

$$= \frac{1}{4} e^u \Big|_0^4$$

$$= \frac{1}{4}(e^4 - e^0)$$

$$= \frac{e^4 - 1}{4}$$

37. **(C)** An efficient method to find the area is shift $h(x) = x^2 - 4$ down 5 units so the entire region lies between the x-axis and the new function, $g(x) = x^2 - 9$.

Region area $= -\int_0^3 x^2 - 9\,dx$ since the entire region lies below the x-axis.

$$-\int_0^3 x^2 - 9\,dx = \int_3^0 x^2 - 9\,dx$$

$$= \frac{1}{3} x^3 - 9x \Big|_3^0$$

$$= 0 - \left(\frac{1}{3} \cdot 27 - 9 \cdot 3\right)$$

$$= 18$$

38. **28** If $h(x) = f(x) \cdot g(x)$, applying product rule,

$$h'(x) = f(x) \cdot g'(x) + g(x) \cdot f'(x)$$
$$h'(2) = f(2) \cdot g'(2) + g(2) \cdot f'(2)$$
$$= 3 \cdot -4 + 8 \cdot 5$$
$$= 28$$

39. **(D)** Base $= 2x$ and Height $= 9 - x^2$

Define area as, $A(x) = \dfrac{1}{2}(2x)(9 - x^2)$.

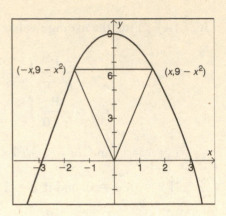

$$A(x) = 9x - x^3$$
$$A'(x) = 9 - 3x^2$$
$$9 - 3x^2 = 0$$
$$x^2 = 3$$
$$x = \sqrt{3}$$

$$A(\sqrt{3}) = 9\sqrt{3} - (\sqrt{3})^3$$
$$= 9\sqrt{3} - 3\sqrt{3}$$
$$= 6\sqrt{3}$$

40. **(D)** For $0 < t < 1$, $v(t) < 0$, and for $1 < t < 3$, $v(t) > 0$.

$$\text{Total distance } = \int_1^3 t \cdot (t^2 - 1)^{\left(\frac{1}{3}\right)} dt - \int_0^1 t \cdot (t^2 - 1)^{\left(\frac{1}{3}\right)} dt$$

$$= \frac{3}{8}(t^2 - 1)^{\left(\frac{4}{3}\right)}\Big|_1^3 - \frac{3}{8}(t^2 - 1)^{\left(\frac{4}{3}\right)}\Big|_0^1$$

$$= \frac{3}{8}[(8)^{\left(\frac{4}{3}\right)} - 0] - \frac{3}{8}[0 - (-1)^{\left(\frac{4}{3}\right)}]$$

$$= \frac{3}{8} \cdot 16 + \frac{3}{8} \cdot 1$$

$$= \frac{51}{8}$$

41. **(B)** The slope of the line tangent to a curve is the value of the derivative at that point.

$$f(x) = \sin(x^3 + x)$$

Apply chain rule.

$$f'(x)\big|_{x=0} = \cos(x^3 + x) \cdot d(x^3 + x)\big|_{x=0}$$
$$= \cos(x^3 + x) \cdot (3x^2 + 1)\big|_{x=0}$$
$$= \cos(0) \cdot (0 + 1)$$
$$= 1$$

42. **(A)** h is decreasing when $h'(x) < 0$.

If $h'(x) = x(x-2)^2 (x+1)$, then $h'(x) = 0$ when $x = 0$, $x = 2$, or $x = -1$.

Test x values around the zeros of $h'(x)$ to determine where h' is negative.

x	$h'(x)$
-2	$(-2)(16)(-1) > 0$
-1	0
$\dfrac{-1}{2}$	$\left(\dfrac{-1}{2}\right)\left(\dfrac{25}{4}\right)\left(\dfrac{1}{2}\right) < 0$
0	0
1	$(1)(1)(2) > 0$
2	0
3	$(3)(1)(4) > 0$

h is decreasing on the interval $[-1, 0]$ since $h'(x) < 0$.

43. **(C)** $\dfrac{dA}{dt} = 0.4A$ so $\dfrac{dA}{A} = 0.4dt$

$$\int \frac{dA}{A} = \int 0.4\, dt$$
$$\ln |A| = 0.4t + C$$
$$|A| = e^{0.4t+C}$$
$$|A| = e^c e^{0.4t}$$

Using $A(0) = 12000$,
$$|12000| = e^C e^0 \Rightarrow 12000 = e^c$$
$$A = 12000e^{0.4t}$$

44. **(A)** The particle is farthest right when $x(t) = 12t - t^3$ is a maximum.
$$x'(t) = 12 - 3t^2 \text{ and } t \geq 0$$
$$12 - 3t^2 = 0 \Rightarrow t = 2$$

Acceleration is $x''(t) = -6t$. Since this is negative for all $t > 0$, $t = 2$ must be a maximum.
$$x''(2) = -12$$

PRACTICE TEST 2

CLEP Calculus

Also available at the REA Study Center (*www.rea.com/studycenter*)

This practice test is also offered online at the REA Study Center. All CLEP exams are administered on computer, and our test is formatted to simulate test-day conditions. We recommend that you take the online version of this test to receive these added benefits:

- **Timed testing conditions** – helps you gauge how much time you can spend on each question
- **Automatic scoring** – find out how you did on the test, instantly
- **On-screen detailed explanations of answers** – gives you the correct answer and explains why the other answer choices are wrong
- **Diagnostic score reports** – pinpoint where you're strongest and where you need to focus your study

PRACTICE TEST 2

CLEP Calculus

(Answer sheets appear in the back of the book.)

TOTAL TIME: 90 Minutes
44 Questions

Section I

TIME: 50 Minutes
27 Questions

Directions: Solve each of the following problems without using a calculator. Choose the best answer from those provided. Some questions will require you to enter a numerical answer in the box provided.

Notes: (1) Figures that accompany questions are intended to provide information useful in answering the questions. All figures lie in a plane unless otherwise indicated. The figures are drawn as accurately as possible EXCEPT when it is stated in a specific question that the figure is not drawn to scale. Straight lines and smooth curves may appear slightly jagged.

(2) Unless otherwise specified, all angles are measured in radians, and all numbers used are real numbers.

(3) Unless otherwise specified, the domain of any function f is assumed to be the set of all real numbers x for which $f(x)$ is a real number. The range of f is assumed to be the set of all real numbers $f(x)$, where x is the domain of f.

(4) In this exam, ln x denotes the natural logarithm of x (the logarithm to the base e).

(5) The inverse of a trigonometric function f may be indicated using the inverse function notation f^{-1} or with the prefix "arc" (e.g., $\sin^{-1} x = \arcsin x$).

1. Which of the following is equivalent to $\lim\limits_{x\to 4}\left(\dfrac{3x^2+6x}{x-1}\right)$?

    I. $\lim\limits_{x\to 4}(3x)\cdot\lim\limits_{x\to 4}\left(\dfrac{x+2}{x-1}\right)$

    II. $(3x)\lim\limits_{x\to 4}\left(\dfrac{x+2}{x-1}\right)$

    III. $\dfrac{\lim\limits_{x\to 4}(3x^2+6x)}{\lim\limits_{x\to 4}(x-1)}$

    (A) I only

    (B) I and II

    (C) I and III

    (D) II and III

    (E) III only

2. $\displaystyle\int (x+3)^4\,dx =$

    (A) $4(x+3)+C$

    (B) $\dfrac{1}{5}x^5+81x+C$

    (C) $\dfrac{1}{4}(x+3)^5+C$

    (D) $\dfrac{1}{5}x(x+3)^5+C$

    (E) $\dfrac{1}{5}(x+3)^5+C$

3. If $y = e^x$, then $\dfrac{dy}{dx}$ evaluated at $x = 3$ is:

    (A) e

    (B) $\dfrac{e^4}{4}$

    (C) e^3

    (D) $3e^2$

    (E) $3e^3$

4. If $y = \dfrac{x^3}{e^x}$, then $\dfrac{dy}{dx} =$

(A) $\dfrac{3x^2 - x^3}{e^x}$

(B) $\dfrac{3x^2 + x^3}{e^x}$

(C) $\dfrac{3x^2 e^x - x^4 e^{(x-1)}}{e^{(2x)}}$

(D) $\dfrac{3x^2}{e^x}$

(E) $\dfrac{3x^2}{xe^{(x-1)}}$

5. $\displaystyle\int \dfrac{\cos^3(2x)}{\sin^2(2x) - 1} dx =$

(A) $\dfrac{-1}{2}\sin(2x) + C$

(B) $2\sin(2x) + C$

(C) $-\sin(2x) + C$

(D) $\dfrac{3\cos^4(2x)}{4\sin^3(2x) - x} + C$

(E) $\dfrac{1}{8}\cos^4(2x) + C$

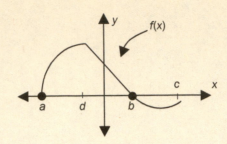

6. The graph of $f(x)$ is shown above. Which graph below could be the graph of $f'(x)$?

(A)

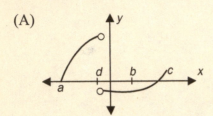

(B)

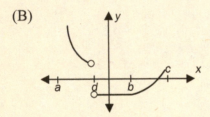

(C)

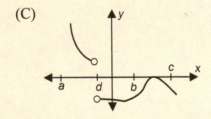

(D)

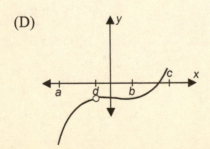

(E)

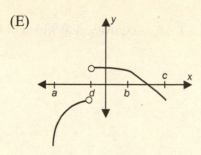

7. If $g(x) = x \cdot \ln(3x)$, find $\dfrac{d^2 g}{dx^2}$.

(A) $\dfrac{-1}{x^2}$

(B) $\dfrac{1}{3x}$

(C) 0

(D) $\dfrac{1}{x}$

(E) $\dfrac{3}{x}$

8. Let $f(x) = \begin{cases} x+3 & \text{for } x < 2 \\ 4 & \text{for } x = 2 \\ 10 - x^2 & \text{for } x > 2 \end{cases}$

What is $\lim\limits_{x \to 2} f(x)$?

(A) 2

(B) 4

(C) 5

(D) 6

(E) The limit does not exist.

9. If $f(x) = x^2 \cos(x)$, then $f'(x) =$

(A) $2x \cdot \sin(x)$

(B) $-2x \cdot \sin(x)$

(C) $2x - \sin(x)$

(D) $2x \cdot \cos(x) - x^2 \cdot \sin(x)$

(E) $2x \cdot \cos(x) + x^2 \cdot \sin(x)$

10. For x in the closed interval $[0, 3]$, $\lim\limits_{n \to \infty} \sum\limits_{i=1}^{n} (x_i)^2 \Delta x_i$ is equivalent to which of the following?

I. ∞

II. $\int_0^3 x^2 \, dx$

III. $\int_1^\infty x^2 \, dx$

(A) I only

(B) II only

(C) III only

(D) I and III

(E) I, II and III

11. If $y = \arctan(u)$ and $u = \sin(x)$, find $\dfrac{dy}{dx}$.

(A) $\dfrac{1}{1 + \sin^2(x)}$

(B) $\dfrac{\cos(x)}{1 + \sin^2(x)}$

(C) $\dfrac{1}{\cos(x)}$

(D) $\sec^2[\sin(x)] \cdot \cos(x)$

(E) $\sec^2[\sin(x)]$

12. Let f be a continuous function for all real numbers with an inflection point at $x = a$. Which statement MUST be true?

(A) $f''(a) = 0$.

(B) $f'(a)$ is continuous at $x = a$.

(C) $f'(a)$ has a local extremum at $x = a$.

(D) f has a tangent line at $x = a$.

(E) $f''(x) < 0$ for $x < a$.

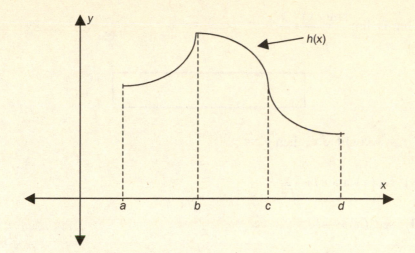

13. The graph of $h(x)$ is shown above. On which of the following intervals are $\dfrac{dh}{dx} < 0$ and $\dfrac{d^2h}{dx^2} > 0$?

   I. $a < x < b$
  II. $b < x < c$
 III. $c < x < d$

(A) I only

(B) II only

(C) III only

(D) I and III

(E) II and III

14. The derivative of the function h is defined on the interval $[0, b]$. $\displaystyle\int_0^b h'(x)\,dx =$

(A) $h(b)$

(B) $h(b) - h(0)$

(C) $|h(b) - h(0)|$

(D) $|h(b)| - |h(0)|$

(E) $h''(b) - h''(0)$

15. What is $\lim\limits_{x\to 0} \dfrac{\sin(2x)+x}{x}$?

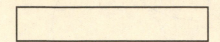

16. If $y = \sec(x) + \sqrt{x}$, then $y' =$

   (A) $\sec(x)\tan(x) + \dfrac{1}{\sqrt{x}}$

   (B) $\sec^2(x) + \dfrac{1}{\sqrt{x}}$

   (C) $\csc(x) + \dfrac{1}{2\sqrt{x}}$

   (D) $\sec(x)\tan(x) + \dfrac{1}{2\sqrt{x}}$

   (E) $\sqrt{x}\cdot\sec(x)\tan(x) + \dfrac{\sec(x)}{2\sqrt{x}}$

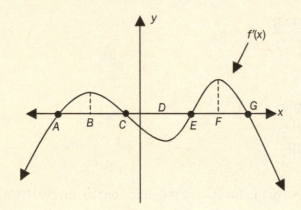

17. The graph of $f'(x)$ is shown above. At which x-values does the graph of f have a local maximum?

   (A) C and G

   (B) A and E

   (C) B and F

   (D) B, D and F

   (E) A, C, E and G

18. The graph of $y = \ln(1 + x^2)$ is concave up for

    (A) $(-\infty, -1)$ and $(1, \infty)$
    (B) $(0, 1)$
    (C) $(-1, 1)$
    (D) All real numbers
    (E) No real numbers

19. From $t = 0$ to 3 seconds, the acceleration of a particle is $2\,\dfrac{\text{ft}}{\text{s}^2}$.
    From $t = 3$ to 6 seconds, its acceleration decreases linearly until the acceleration is $0\,\dfrac{\text{ft}}{\text{s}^2}$. If the particle's initial velocity was $5\,\dfrac{\text{ft}}{\text{s}}$, its velocity at $t = 6$ seconds is

    (A) $0\,\dfrac{\text{ft}}{\text{s}}$

    (B) $3\,\dfrac{\text{ft}}{\text{s}}$

    (C) $8\,\dfrac{\text{ft}}{\text{s}}$

    (D) $9\,\dfrac{\text{ft}}{\text{s}}$

    (E) $14\,\dfrac{\text{ft}}{\text{s}}$

20. $\displaystyle\lim_{x\to\frac{3}{2}}\left(\dfrac{\cos(\pi x) - \cos\left(\frac{3\pi}{2}\right)}{x - \frac{3}{2}}\right) =$

    (A) $-\pi$
    (B) -1
    (C) 0
    (D) 1
    (E) π

21. If $h(x) = \begin{cases} 3x+1 & \text{for } x > 2 \\ \frac{3}{4}x^2+2 & \text{for } x \leq 2 \end{cases}$, at $x = 2$ $h(x)$ is

    (A) both continuous and differentiable.
    (B) neither continuous nor differentiable.
    (C) continuous but not differentiable.
    (D) differentiable but not continuous.
    (E) asymptotic to the line $x = 2$.

22. On which intervals is the graph of $y = x(x - 3)^2$ increasing?

    (A) $[3, \infty)$
    (B) $[1, 3]$
    (C) $[0, \infty)$
    (D) $(-\infty, 0]$ or $[1, \infty)$
    (E) $(-\infty, 1]$ or $[3, \infty)$

23. Which of the following conditions guarantees that a left-oriented rectangular approximation of the area under a positive function on an interval $[a, b]$ is always an underestimate of the true area?

    (A) The function is increasing on $[a, b]$.
    (B) The function is decreasing on $[a, b]$.
    (C) The function is concave up on $[a, b]$.
    (D) The function is concave down on $[a, b]$.
    (E) The function has a local maximum in (a, b).

24. Let $f(x) = x + \sqrt{x+1}$ and $g(x)$ be the inverse of f. Find $g'(5)$.

    (A) $\dfrac{-5}{4}$

    (B) $\dfrac{-4}{5}$

    (C) $\dfrac{4}{5}$

(D) $1 + \dfrac{1}{2\sqrt{6}}$

(E) $\dfrac{5}{4}$

25. Let $h(x)$ be a continuous function on the interval $[1, 5]$. If $2 \leqslant h(x) \leqslant 6$, then what is the greatest possible value of $\displaystyle\int_1^5 2h(x)\,dx$?

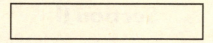

26.

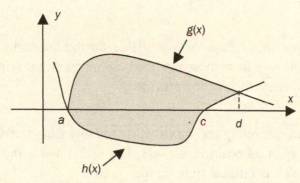

Which of the following integral expressions represents the area of the shaded region above?

(A) $\displaystyle\int_a^d (g(x) - h(x))\,dx$

(B) $\displaystyle\int_a^d (g(x) - |\,h(x)\,|)\,dx$

(C) $\displaystyle\int_a^d (g(x) + h(x))\,dx$

(D) $\displaystyle\int_a^d g(x)\,dx - \int_a^c h(x)\,dx$

(E) $\displaystyle\int_a^c (g(x) - |\,h(x)\,|)\,dx + \int_c^d (g(x) - h(x))\,dx$

27. The instantaneous rate of change of $f(x)$ is directly proportional to $f(x)$. Which of the following could be $f(x)$?

    (A) $e^{(4x-1)}$

    (B) $2x + 9$

    (C) $\sqrt{5x+7}$

    (D) $\ln(3x^2 - 1)$

    (E) $\sin(8x)$

Section II

TIME: 40 Minutes
        17 Questions

Directions: A graphing calculator is available for the questions in this section. Choose the best answer from those provided. Some questions will require you to enter a numerical answer in the box provided.

28. A particle moves along the x-axis so that at any time $t \geq 0$ its velocity is $v(t) = 9 - t^2$. If its position at $t = 1$, is $x = 2$, what is the position of the particle when it is farthest right on the x-axis?

    (A) $9\dfrac{1}{3}$

    (B) $11\dfrac{1}{3}$

    (C) 18

    (D) 20

    (E) $28\dfrac{2}{3}$

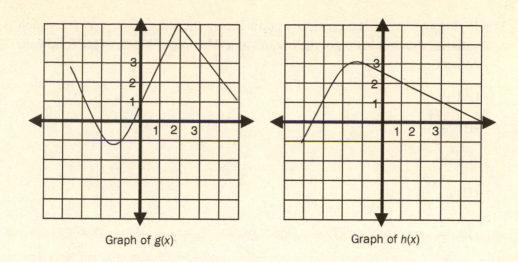

Graph of g(x)   Graph of h(x)

29. The graphs of $g(x)$ and $h(x)$ are shown above. If $f(x) = g(x) \cdot h(x)$, use the graphs to choose the best estimate of the rate of change of f at $x = 1$.

(A) $\dfrac{-5}{2}$

(B) -1

(C) $\dfrac{5}{2}$

(D) 5

(E) $\dfrac{11}{2}$

30. Find the slope of the line tangent to $3xy + y^2 + 4x = 8$ at the point $(1, 1)$.

(A) -3

(B) -2

(C) $\dfrac{-7}{5}$

(D) $\dfrac{-4}{5}$

(E) $\dfrac{1}{5}$

31. If $f(x) = x^3 - 1$, there exists a number c in the interval $0 < x < 2$ which satisfies the conclusion of the Mean Value Theorem. Which of the following could be c?

(A) 0

(B) $\dfrac{\sqrt{3}}{3}$

(C) 1

(D) $\dfrac{2\sqrt{3}}{3}$

(E) $\dfrac{4}{3}$

32. If $\displaystyle\int_1^9 \left(k - \dfrac{1}{\sqrt{x}} \right) dx = 60$, then $k =$

(A) 4

(B) 5

(C) 6

(D) 7

(E) 8

33. The medicine in a person's system has a rate of decrease in direct proportion to the amount present in a person's bloodstream at any given time. If a 50 mg dose decreases to 20 mg in 2 hours, the constant of proportionality is

(A) -30

(B) -15

(C) $\dfrac{1}{2} \ln\left(\dfrac{2}{5}\right)$

(D) $\dfrac{-2}{5}$

(E) $e^{-0.4}$

$g(4)$	$g'(4)$	$h(4)$	$h'(4)$
8	6	2	-3

34. Given $f(x) = \dfrac{g(x)}{h(x)}$, use the table of values above to find $f'(4)$.

    (A) 36
    (B) 18
    (C) 9
    (D) -2
    (E) -3

35. An open-top rectangular box with square base and a volume of 3 cubic feet must be constructed. The material for the base costs \$.06 per square foot, and the material for the sides costs \$.08 per square foot. The minimum cost to construct the box is

    (A) \$.64
    (B) \$.72
    (C) \$.86
    (D) \$1.02
    (E) \$2.00

36. Using the trapezoidal rule, with 3 trapezoids, determine the <u>first quadrant</u> area under $f(x) = 18 - 3x - x^2$.

37. Let R be the first quadrant region enclosed by $y = e^2$, $y = e^x$ and the y-axis. The area of region R is

(A) $\dfrac{e^3}{2} - e^2$

(B) $e^2 - 1$

(C) e^2

(D) $e^2 + 1$

(E) $e + 2e^2 - \dfrac{e^3}{3}$

38. Using the substitution $u = 5x^2 - 2$ to rewrite the integral and limits of $\displaystyle\int_1^2 \dfrac{x \, dx}{\sqrt{5x^2 - 2}}$, the resulting integral becomes

(A) $\displaystyle\int_3^{18} \dfrac{1}{5} u^{\left(\frac{1}{2}\right)} du$

(B) $\displaystyle\int_3^{18} \dfrac{1}{5} u^{\left(\frac{-1}{2}\right)} du$

(C) $\displaystyle\int_3^{18} \dfrac{1}{10} u^{\left(\frac{-1}{2}\right)} du$

(D) $\displaystyle\int_3^{18} \dfrac{1}{10} u^{\left(\frac{1}{2}\right)} du$

(E) $\displaystyle\int_1^2 \dfrac{1}{5} u^{\left(\frac{-1}{2}\right)} du$

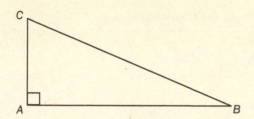

39. In right triangle ABC above, at the instant $AC = 3$ and $AB = 6$, leg $\overline{AC}$ is increasing at a rate of 2 inches per second and leg $\overline{AB}$ is increasing at 8 inches per second. In square inches per second, how fast is the area of the triangle increasing?

    (A) 36

    (B) 18

    (C) 16

    (D) 12

    (E) 8

40. Let $h(x) = x^3 - 3x$. Find the average rate of change of $h(x)$ on the interval $[1, 4]$.

    (A) 15

    (B) $\dfrac{50}{3}$

    (C) 18

    (D) $\dfrac{45}{2}$

    (E) 25

41. Find the slope of the curve $k(x) = \sin^2(4x)$ at $x = \dfrac{\pi}{6}$.

    (A) $-2\sqrt{3}$

    (B) -1

    (C) $\dfrac{-\sqrt{3}}{2}$

    (D) $\sqrt{3}$

    (E) $4\sqrt{3}$

42. Find the equation of the line tangent to $y = -(1 - 2x)^4$ at the point $(1, -1)$.

    (A) $y = -8x + 7$
    (B) $y = 8x - 9$
    (C) $y = 4x - 5$
    (D) $y = -4x + 3$
    (E) $y = -8x + 9$

43. The position of a particle moving along the x-axis as a function of time, t, is given by $x(t) = \frac{1}{6}t^3 - t^2 + 3t - 1$ for $t \geqslant 0$. The particle's velocity becomes three times its initial velocity when $t =$

    (A) 1
    (B) 2
    (C) 4
    (D) 6
    (E) 10

44. For $0 \leqslant x \leqslant 2$, the average distance of a point on the graph of $y = x^2$ to the line $x = 2$ is

    (A) $\dfrac{1}{3}$

    (B) $\dfrac{1}{2}$

    (C) $\dfrac{2}{3}$

    (D) 1

    (E) $\dfrac{4}{3}$

PRACTICE TEST 2

Answer Key

1. (C)	16. (D)	31. (D)
2. (E)	17. (A)	32. (E)
3. (C)	18. (C)	33. (C)
4. (A)	19. (E)	34. (C)
5. (A)	20. (E)	35. (B)
6. (B)	21. (B)	36. 31
7. (D)	22. (E)	37. (D)
8. (E)	23. (A)	38. (C)
9. (D)	24. (C)	39. (B)
10. (B)	25. 48	40. (C)
11. (B)	26. (A)	41. (A)
12. (D)	27. (A)	42. (A)
13. (C)	28. (B)	43. (D)
14. (B)	29. (C)	44. (C)
15. 3	30. (C)	

PRACTICE TEST 2

Detailed Explanations of Answers

1. **(C)** Since the limit exists for both the numerator and denominator, properties of limits allow writing the fraction as a product of the limits of each factor (choice I) or as a quotient (choice III) of each limit. One is never allowed to factor the independent variable outside of the limit expression (choice II).

2. **(E)** Use the integral form, $\int u^n \, du = \dfrac{u^{n+1}}{n+1} + C$

   Let $u = x + 3$, and $du = dx$.

   $$\int (x+3)^4 \, dx = \int u^4 \, du$$

   $$= \frac{u^5}{5} + C$$

   $$= \frac{(x+3)^5}{5} + C$$

3. **(C)** If $y = e^x$, then $\dfrac{dy}{dx} = e^x$ and when $x = 3$, $\dfrac{dy}{dx} = e^3$

4. **(A)** If $y = \dfrac{x^3}{e^x}$, apply the quotient rule to find $\dfrac{dy}{dx}$.

   $$\frac{dy}{dx} = \frac{e^x \cdot \frac{d}{dx}(x^3) - x^3 \cdot \frac{d}{dx}(e^x)}{(e^x)^2}$$

   $$= \frac{e^x \cdot 3x^2 - x^3 \cdot e^x}{e^x \cdot e^x}$$

   $$= \frac{e^x(3x^2 - x^3)}{e^x \cdot e^x}$$

   $$= \frac{3x^2 - x^3}{e^x}$$

5. **(A)** Use a trigonometric identity to simplify the integrand before antidifferentiation.

$$1 - \sin^2(2x) = \cos^2(2x), \text{ so } \sin^2(2x) - 1 = -\cos^2(2x).$$

$$\frac{\cos^3(2x)}{\sin^2(2x)-1} = \frac{\cos^3(2x)}{-\cos^2(2x)}$$

$$= -\cos(2x)$$

$$\int \frac{\cos^3(2x)}{\sin^2(2x)-1}\, dx = \int -\cos(2x)\, dx$$

Let $u = 2x$, so $du = 2dx$, and $dx = \frac{1}{2}\, du$

$$\int -\cos(2x)\, dx = \frac{-1}{2} \int \cos(u)\, du$$

$$= \frac{-1}{2} \sin(u) + C$$

$$= \frac{-1}{2} \sin(2x) + C$$

6. **(B)**

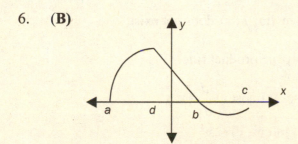

Imagine a tangent line on the given graph, at $x = a$. The tangent starts with a very large slope. As it moves toward the corner between a and b at $x = d$, the slope of the tangent line stays positive, but begins to decrease. At $x = d$ the derivative is undefined, so the graph of $f'(x)$ has an open circle. At this point the only choices with a positive, decreasing derivative are (B) or (C). Between $x = d$ to the local minimum of f, $f(x)$ is decreasing, so $f'(x)$ is negative. At the minimum, $f'(x) = 0$, but to the right of the minimum, $f(x)$ increases, so $f'(x) > 0$. On graph (B), $f'(x)$ goes from negative to zero to positive in this interval, but graph (C) does not.

7. **(D)** If $g(x) = x \cdot \ln(3x)$, use the product rule.

$$\frac{dg}{dx} = x \cdot \frac{d}{dx} \ln(3x) + \ln(3x) \cdot \frac{d}{dx}(x)$$

$$= x \cdot \frac{3}{3x} + \ln(3x)$$

$$= 1 + \ln(3x)$$

$$\frac{d^2g}{dx^2} = \frac{d}{dx}(1 + \ln(3x))$$

$$= 0 + \frac{1}{3x} \cdot \frac{d}{dx}(3x)$$

$$= \frac{1}{3x} \cdot 3$$

$$= \frac{1}{x}$$

8. **(E)** In order for $\lim_{x \to 2} f(x)$ to exist, the left and right-handed limits must be equal. $\lim_{x \to 2^-} f(x) = 2 + 3 = 5$ and $\lim_{x \to 2^+} f(x) = 10 - 2^2 = 6$. Since the limits from both sides are different $\lim_{x \to 2} f(x)$ does not exist.

9. **(D)** If $f(x) = x^2 \cos(x)$, apply the product rule.

$$f'(x) = x^2 \frac{d}{dx}(\cos(x)) + \cos(x) \frac{d}{dx}(x^2)$$

$$= x^2 \cdot -\sin(x) + \cos(x) \cdot 2x$$

$$= 2x \cdot \cos(x) - x^2 \cdot \sin(x)$$

10. **(B)** In combination with the given interval [0, 3], the summation expression is the Riemann sum for calculating the area under $f(x) = x^2$, so

$$\lim_{n \to \infty} \sum_{i=1}^{n} (x_i)^2 \Delta x_i = \int_0^3 x^2 dx .$$

$i = 1$ to $i = n$ numbers each of the n rectangles.

$(x_i)^2$ is the value of the function for x in the ith subinterval, the height of each rectangle.

Δx_i is the width of any i^{th} rectangle.

$\lim_{n \to \infty}$ finds the sum of the areas as the number of rectangles increases infinitely.

11. **(B)** If $y = \arctan(u)$ and $u = \sin(x)$, use the chain rule $\dfrac{dy}{dx} = \dfrac{dy}{du} \cdot \dfrac{du}{dx}$.

$$\frac{dy}{du} = \frac{1}{1+u^2}$$

$$\frac{du}{dx} = \cos(x)$$

$$\frac{dy}{dx} = \frac{dy}{du} \cdot \frac{du}{dx}$$

$$= \frac{1}{1+u^2} \cdot \cos(x)$$

$$= \frac{\cos(x)}{1+\sin^2(x)}$$

12. **(D)** Anywhere a function has an inflection point, the function must be continuous and have a change in its concavity. A tangent line can always be drawn at that point. The function $y = \sqrt[3]{x}$ is a counterexample for choices (A), (B), (C), and (E). At the inflection point at $x = 0$, $f'(x)$ and $f''(x)$ are both undefined. For $f < 0$, $f''(x) > 0$.

13. **(C)** If $\dfrac{dh}{dx} < 0$, h must be decreasing. If $\dfrac{d^2h}{dx^2} > 0$, h must be concave up. The only place the given graph is decreasing and concave up is $c < x < d$.

14. **(B)** Use the Fundamental Theorem of Calculus for integral evaluation.

$$\int_0^b h'(x)\, dx = h(x)\big|_0^b$$

$$= h(b) - h(0)$$

15. **3** $\displaystyle\lim_{x \to 0} \frac{\sin(2x)+x}{x} = \lim_{x \to 0} \frac{\sin(2x)}{x} + \lim_{x \to 0} \frac{x}{x}$

To determine $\displaystyle\lim_{x \to 0} \frac{\sin(2x)}{x}$, use L'Hôpital's Rule, $\displaystyle\lim_{x \to 0} \frac{f(x)}{g(x)} = \lim_{x \to 0} \frac{f'(x)}{g'(x)}$ whenever both $\displaystyle\lim_{x \to 0} f(x)$ and $\displaystyle\lim_{x \to 0} g(x)$ are either zero or infinity.

In this example, each of these limits is zero.

$$f(x) = \sin 2x, \text{ so } f'(x) = 2 \cos 2x$$

$$g(x) = x, \text{ so } g'(x) = 1.$$

Thus, $\lim\limits_{x\to 0}\dfrac{\sin(2x)}{x}=\lim\limits_{x\to 0}\dfrac{2\cos 2x}{1}=2\cos 0=2$.

We know that $\lim\limits_{x\to 0}\dfrac{x}{x}=\lim\limits_{x\to 0}1=1$.

Therefore $\lim\limits_{x\to 0}\dfrac{\sin(2x)+x}{x}=2+1=3$.

16. **(D)** Think of $y=\sec(x)+\sqrt{x}$ as $y=\sec(x)+(x)^{\left(\frac{1}{2}\right)}$ and apply the derivative of a sum and of a power.

$$y'=\frac{d}{dx}(\sec(x))+\frac{d}{dx}\left((x)^{\left(\frac{1}{2}\right)}\right)$$

$$y'=\sec(x)\tan(x)+\frac{1}{2}x^{\left(\frac{-1}{2}\right)}$$

$$=\sec(x)\tan(x)+\frac{1}{2\sqrt{x}}$$

17. **(A)** For the graph of f to have a local maximum at a point, $f'(x)$ must go from positive, to zero, to negative as it crosses the x-value where the maximum exists. Remembering that the y-values on the given graph are the values of $f'(x)$, the necessary sign change occurs at points C and G.

18. **(C)** Find where y'' is greater than zero.

$$y'=\frac{2x}{1+x^2}$$

$$y''=\frac{(1+x^2)\cdot 2-2x\cdot 2x}{(1+x^2)^2}$$

$$=\frac{2-2x^2}{(1+x^2)^2}$$

$$\frac{2-2x^2}{(1+x^2)^2}=0 \text{ when } 2-2x^2=0$$

$$x=\pm 1$$

If $x<-1$ or $x>1$, then $y''<0$.

If $-1<x<1$, then $y''>0$, and $y=\ln(1+x^2)$ is concave up.

19. **(E)**

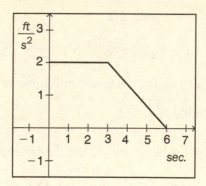

The area under the graph of an acceleration curve accumulates change in velocity. The described acceleration graph is shown above. Treating the region as a rectangle and a triangle, the area is $2\,\frac{\text{ft}}{s^2}\cdot 3s + \frac{1}{2}\cdot 2\,\frac{\text{ft}}{s^2}\cdot 3s = 9\,\frac{\text{ft}}{s}$. Add the initial velocity of $5\,\frac{\text{ft}}{s}$, and the velocity at $t = 6$ seconds is $14\,\frac{\text{ft}}{s}$.

20. **(E)** Compare $\lim\limits_{x\to\frac{3}{2}}\left(\dfrac{\cos(\pi x) - \cos\left(\frac{3\pi}{2}\right)}{x - \frac{3}{2}}\right)$ to the definition of a derivative form

$f'(a) = \lim\limits_{x\to a}\dfrac{f(x) - f(a)}{x - a}$. For the given limit, $f(x) = \cos(\pi x)$ and $a = \dfrac{3}{2}$.

$$\frac{d}{dx}[\cos(\pi x)]\bigg|_{x=\frac{3}{2}} = -\pi\cdot[\sin(\pi x)]\bigg|_{x=\frac{3}{2}}$$

$$= -\pi\cdot\sin\left(\frac{3\pi}{2}\right)$$

$$= \pi$$

Alternatively, since the limit is an indeterminate form, $\dfrac{0}{0}$, L'Hôpital's rule may also be applied.

$$\lim_{x\to\frac{3}{2}}\left(\frac{\cos(\pi x) - \cos\left(\frac{3\pi}{2}\right)}{x - \frac{3}{2}}\right) = \lim_{x\to\frac{3}{2}}\left(\frac{-\pi\sin(\pi x) - 0}{1 - 0}\right)$$

$$= -\pi\sin\left(\frac{3\pi}{2}\right)$$

$$= \pi$$

21. **(B)** $h(x) = \begin{cases} 3x+1 & \text{for } x > 2 \\ \frac{3}{4}x^2 + 2 & \text{for } x \le 2 \end{cases}$

$\lim\limits_{x \to 2^+} h(x) = 3 \cdot 2 + 1 = 7$ and $\lim\limits_{x \to 2^-} h(x) = \frac{3}{4} \cdot 2^2 + 2 = 5$.

$\lim\limits_{x \to 2^+} h(x) \ne \lim\limits_{x \to 2^-} h(x)$, so $h(x)$ is not continuous at $x = 2$, therefore $h(x)$ cannot be differentiable at $x = 2$.

22. **(E)** Use the product rule to find y', and determine where it is positive.

$$y' = x \cdot \frac{d}{dx}(x-3)^2 + (x-3)^2 \cdot \frac{d}{dx}(x)$$
$$= x \cdot 2(x-3) + (x-3)^2 \cdot 1$$
$$= (x-3)[(2x + (x-3)]$$
$$= (x-3)(3x-3); \quad (x-3)(3x-3) = 0 \Rightarrow x = 3 \text{ or } x = 1$$

If $x < 1$, then $y' > 0$.

If $1 < x < 3$, then $y' < 0$.

If $x > 3$, then $y' > 0$.

y is increasing for $x \le 1$ or $x \ge 3$. In interval notation, $(-\infty, 1]$ or $[3, \infty)$.

Note: Endpoints of intervals of increasing or decreasing may be included or omitted. This practice varies from textbook to textbook.

23. **(A)** Whether a function is concave up or down, if it is increasing, using the left edge of each interval for the height of a rectangle will choose the smallest height in each interval, and therefore underestimate each area. If a function is decreasing or, increasing and decreasing, on an interval, it is possible to overestimate the true area with left-oriented rectangles.

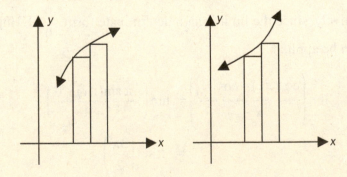

24. **(C)** It is very difficult to find the inverse of $f(x) = x + \sqrt{x+1}$, so use the rule for derivatives of inverse functions, $g'(x) = \dfrac{1}{f'(g(x))}$.

If f and g are inverses, $x = 5$ on g corresponds to $y = 5$ on f.

Solving $f(x) = 5$, leads to $x = 3$. Therefore, $f(3) = 5$ and $g(5) = 3$, so $g'(5) = \dfrac{1}{f'(3)}$.

$$f'(x) = 1 + \frac{1}{2\sqrt{x+1}}$$

$$f'(3) = 1 + \frac{1}{2\sqrt{4}}$$

$$f'(3) = \frac{5}{4}$$

$$g'(5) = \frac{1}{f'(3)} = \frac{4}{5}$$

25. **48** The greatest possible value of $\int_{1}^{5} 2h(x)\,dx$ occurs if $h(x)$ is as large as possible on the entire interval [1, 5].

Let $h(x) = 6$.

$$\int_{1}^{5} 2h(x)\,dx = \int_{1}^{5} 2 \cdot 6\,dx$$

$$= 12x \big|_{1}^{5}$$

$$= 60 - 12$$

$$= 48$$

26. **(A)** Whether or not the lower function has positive or negative values, the height of each rectangle in the Riemann sum is determined by the difference in the y-values, greater function minus lower function. Choice B would make the difference between the positive values on $g(x)$ and the negative values on $h(x)$ into a difference of positive values, thereby making the heights of all rectangles from a to c too small. The correct choice is (A) because $\displaystyle\lim_{n \to \infty} \sum_{i=1}^{n} (g(x_i) - h(x_i)) \cdot \Delta x_i = \int_{a}^{d} (g(x) - h(x))\,dx$.

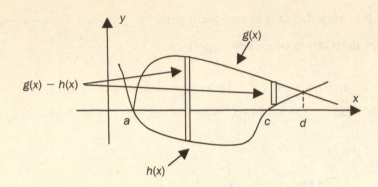

27. **(A)** If the instantaneous rate of change of $f(x)$ is directly proportional to $f(x)$, $f'(x) = k \cdot f(x)$ where k is a constant. Often, this involves powers of e. If $f(x) = e^{(4x-1)}$,

$$\frac{d}{dx}[e^{(4x-1)}] = e^{(4x-1)} \cdot \frac{d}{dx}(4x-1)$$
$$= 4e^{(4x-1)}$$
$$= 4f(x)$$

28. **(B)** When the particle is farthest right, its velocity is going from positive, to zero, to negative. For $v(t) = 9 - t^2$, this happens when $t = 3$ seconds. Integrating velocity from $t = 1$ to $t = 3$ finds the change in position. Add this to the initial position at $t = 1$ second.

$$2 + \int_1^3 9 - t^2 \, dt = 2 + \left(9t - \frac{t^3}{3}\right)\Big|_{t=1}^{t=3}$$
$$= 2 + (27-9) - \left(9 - \frac{1}{3}\right)$$
$$= 11\frac{1}{3}$$

29. **(C)** The rate of change of f at $x = 1$ is $f'(1)$. Applying the product rule,
$$f'(x) = g(x) \cdot h'(x) + h(x) \cdot g'(x).$$
$$f'(1) = g(1) \cdot h'(1) + h(1) \cdot g'(1)$$

Read from the graphs to get function values $h(1)$ and $g(1)$, and estimate slopes for $h'(1)$ and $g'(1)$.

$$f'(1) = g(1) \cdot h'(1) + h(1) \cdot g'(1)$$

$$\approx 3 \cdot \frac{-1}{2} + 2 \cdot 2$$

$$\approx \frac{5}{2}$$

30. **(C)** Use implicit differentiation and remember the product rule on the $3xy$ term.

$$\frac{d}{dx}[3x \cdot y + y^2 + 4x] = \frac{d}{dx}[8]$$

$$3x\frac{dy}{dx} + 3y + 2y\frac{dy}{dx} + 4 = 0$$

Substitute (1, 1) and solve.

$$3 \cdot 1\frac{dy}{dx} + 3 \cdot 1 + 2 \cdot 1\frac{dy}{dx} + 4 = 0$$

$$5\frac{dy}{dx} + 7 = 0$$

$$\frac{dy}{dx} = \frac{-7}{5}$$

31. **(D)**

If $f(x) = x^3 - 1$, $f'(x) = 3x^2$.

Solve $f'(c) = \dfrac{f(2) - f(0)}{2 - 0}$.

$$3c^2 = \frac{7 - (-1)}{2 - 0}$$

$$3c^2 = 4$$

$$c^2 = \frac{4}{3}$$

$$c = \pm\frac{2}{\sqrt{3}}$$

Choose the solution in the interval, and rationalize to $c = \dfrac{2\sqrt{3}}{3}$.

32. **(E)**

$$\int_1^9 \left(k - \frac{1}{\sqrt{x}} \right) dx = 60$$

$$\int_1^9 \left(k - \frac{1}{\sqrt{x}} \right) dx = \left(kx - 2\sqrt{x} \right) \Big|_1^9$$

$$= \left(9k - 2\sqrt{9} \right) - \left(k - 2\sqrt{1} \right)$$

$$= 8k - 4$$

$$8k - 4 = 60$$

$$8k = 64$$

$$k = 8$$

33. **(C)** Let m = the amount of medicine in the bloodstream at any time, t = time, and k = the constant of proportionality.

Let $m(0) = 50$ and $m(2) = 20$.

$$\frac{dm}{dt} = km$$

$$\frac{dm}{m} = k \cdot dt$$

$$\int \frac{dm}{m} = \int k \cdot dt$$

$$\ln(m) = k \cdot t + C$$

$$m = m_0 \cdot e^{kt}$$

$$20 = 50e^{2k}$$

$$\frac{2}{5} = e^{2k}$$

$$\ln\left(\frac{2}{5} \right) = 2k$$

$$\frac{1}{2} \ln\left(\frac{2}{5} \right) = k$$

34. **(C)**   Given $f(x) = \dfrac{g(x)}{h(x)}$, apply the quotient rule and substitute values from the table.

$$f'(x) = \frac{h(x) \cdot g'(x) - g(x) \cdot h'(x)}{[h(x)]^2}$$

$$f'(4) = \frac{h(4) \cdot g'(4) - g(4) \cdot h'(4)}{[h(4)]^2}$$

$$= \frac{2 \cdot 6 - 8 \cdot (-3)}{2^2}$$

$$= 9$$

35. **(B)**   Let $x =$ the length, in feet, of a side of the base.

Let $y =$ the height, in feet, of the box.

The volume is 3 cubic feet, so $x^2 y = 3$ and $y = \dfrac{3}{x^2}$.

Cost is \$.06 times the base area plus \$.08 times the area of each of the four sides.

$$C(x) = .06 \cdot x^2 + (.08) \cdot 4xy$$

$$C(x) = .06x^2 + .32x \cdot \frac{3}{x^2}$$

$$C(x) = .06x^2 - .96x^{-1}$$

$$C'(x) = .12x - .96x^{-2}$$

$$= .12x - \frac{.96}{x^2}$$

To minimize cost, set $C'(x)$ equal to zero.

$$.12x - \frac{.96}{x^2} = 0$$

$$.12x^3 - .96 = 0$$

$$x^3 = 8$$

$$x = 2$$

$$C(2) = .06(2)^2 + .96(2)^{-1}$$

$$C(2) = .24 + .48$$

$$= \$0.72$$

36. **31**

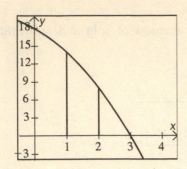

$$A = \frac{h}{2}[f(x_0) + 2f(x_1) + 2f(x_2) + f(x_3)]$$

$$A = \frac{1}{2}[f(0) + 2f(1) + 2f(2) + f(3)]$$

$$= \frac{1}{2}[18 + 2 \cdot 14 + 2 \cdot 8 + 0]$$

$$= 31$$

37. **(D)**

$$e^x = e^2$$

$$x = 2$$

$$\int_0^2 (e^2 - e^x)\, dx = (x \cdot e^2 - e^x)\Big|_0^2$$

$$= (2 \cdot e^2 - e^2) - (0 \cdot e^2 - e^0)$$

$$= (e^2) - (-1)$$

$$= e^2 + 1$$

38. **(C)** Convert the given integral so that both the integrand and limits are in terms of u.

If $u = 5x^2 - 2$, $du = 10x\,dx$ and $\dfrac{1}{10}\,du = x\,dx$.

When $x = 1$, $u = 5 \cdot 1^2 - 2 = 3$. When $x = 2$, $u = 5 \cdot 2^2 - 2 = 18$

Substitute for all parts of the integral.

$$\int_1^2 \frac{x\,dx}{\sqrt{5x^2-2}} = \int_3^{18} \frac{\frac{1}{10}\,du}{\sqrt{u}}$$

$$= \int_3^{18} \frac{1}{10} u^{\left(\frac{-1}{2}\right)}\,du$$

39. **(B)** Let R be the area of triangle ABC. $R = \frac{1}{2}(AC)(AB)$

Differentiate both sides of the equation with respect to t. Use product rule on $(AC)(AB)$.

$$\frac{dR}{dt} = \frac{1}{2}\left[AC \cdot \frac{d}{dt}(AB) + AB \cdot \frac{d}{dt}(AC)\right]$$

$$= \frac{1}{2}[3 \cdot 8 + 6 \cdot 2]$$

$$= 18$$

40. **(C)** $h(x) = x^3 - 3x$

The average rate of change is the slope of the secant joining $(1, h(1))$ to $(4, h(4))$.

$$\frac{h(4) - h(1)}{4 - 1} = \frac{(64 - 12) - (1 - 3)}{3}$$

$$= \frac{52 + 2}{3}$$

$$= 18$$

41. **(A)** $k'(x) = 2\sin(4x) \cdot \frac{d}{dx}(\sin(4x))$

$$= 2\sin(4x)\cos(4x) \cdot \frac{d}{dx}(4x)$$

$$= 2\sin(4x)\cos(4x) \cdot 4$$

$$= 8\sin(4x)\cos(4x)$$

$$k'\left(\frac{\pi}{6}\right) = 8\sin\left(\frac{4\pi}{6}\right)\cos\left(\frac{4\pi}{6}\right)$$

$$= 8 \cdot \frac{\sqrt{3}}{2} \cdot \frac{-1}{2}$$

$$= -2\sqrt{3}$$

42. **(A)** $y = -(1 - 2x)^4$

$$y' = -4(1-2x)^3 \cdot \frac{d}{dx}(1-2x)$$

$$= -4(1-2x)^3(-2)$$

$$= 8(1-2x)^3$$

The slope of the line is, $y'(1) = 8(-1)^3 = -8$.

$$y + 1 = -8(x - 1)$$

$$y + 1 = -8x + 8$$

$$y = -8x + 7$$

43. **(D)** For $x(t) = \frac{1}{6}t^3 - t^2 + 3t - 1$, find t when $v(t) = 3v(0)$.

$$v(t) = x'(t) = \frac{1}{2}t^2 - 2t + 3 \text{ and } v(0) = 3$$

$$\frac{1}{2}t^2 - 2t + 3 = 9$$

$$\frac{1}{2}t^2 - 2t - 6 = 0$$

$$t^2 - 4t - 12 = 0$$

$$(t - 6)(t + 2) = 0$$

$$t = 6 \text{ or } t = -2$$

$$t = 6$$

44. **(C)** For $x \geq 0$, a general point on the graph of $y = x^2$ can be written (x, x^2) or $(\sqrt{y}, y)$. Since the distance is measured horizontally, define each distance as $2 - \sqrt{y}$ and find the average value of the distance function for $0 \leq y \leq 4$.

$$\frac{1}{4-0}\int_0^4 (2-\sqrt{y})\,dy = \frac{1}{4}\left(2y - \frac{2}{3}y^{\left(\frac{3}{2}\right)}\right)\Bigg|_0^4$$

$$= \frac{1}{4}\left(2 \cdot 4 - \frac{2}{3} \cdot 4^{\left(\frac{3}{2}\right)}\right)$$

$$= \frac{1}{4}\left(8 - \frac{16}{3}\right)$$

$$= \frac{2}{3}$$

ANSWER SHEETS

Practice Test 1
Practice Test 2

PRACTICE TEST 1

Answer Sheet

1. Ⓐ Ⓑ Ⓒ Ⓓ Ⓔ
2. Ⓐ Ⓑ Ⓒ Ⓓ Ⓔ
3. [＿＿＿＿＿＿]
4. Ⓐ Ⓑ Ⓒ Ⓓ Ⓔ
5. [＿＿＿＿＿＿]
6. Ⓐ Ⓑ Ⓒ Ⓓ Ⓔ
7. Ⓐ Ⓑ Ⓒ Ⓓ Ⓔ
8. Ⓐ Ⓑ Ⓒ Ⓓ Ⓔ
9. Ⓐ Ⓑ Ⓒ Ⓓ Ⓔ
10. Ⓐ Ⓑ Ⓒ Ⓓ Ⓔ
11. Ⓐ Ⓑ Ⓒ Ⓓ Ⓔ
12. Ⓐ Ⓑ Ⓒ Ⓓ Ⓔ
13. Ⓐ Ⓑ Ⓒ Ⓓ Ⓔ
14. Ⓐ Ⓑ Ⓒ Ⓓ Ⓔ
15. Ⓐ Ⓑ Ⓒ Ⓓ Ⓔ

16. Ⓐ Ⓑ Ⓒ Ⓓ Ⓔ
17. Ⓐ Ⓑ Ⓒ Ⓓ Ⓔ
18. Ⓐ Ⓑ Ⓒ Ⓓ Ⓔ
19. Ⓐ Ⓑ Ⓒ Ⓓ Ⓔ
20. Ⓐ Ⓑ Ⓒ Ⓓ Ⓔ
21. Ⓐ Ⓑ Ⓒ Ⓓ Ⓔ
22. Ⓐ Ⓑ Ⓒ Ⓓ Ⓔ
23. Ⓐ Ⓑ Ⓒ Ⓓ Ⓔ
24. Ⓐ Ⓑ Ⓒ Ⓓ Ⓔ
25. Ⓐ Ⓑ Ⓒ Ⓓ Ⓔ
26. Ⓐ Ⓑ Ⓒ Ⓓ Ⓔ
27. Ⓐ Ⓑ Ⓒ Ⓓ Ⓔ
28. Ⓐ Ⓑ Ⓒ Ⓓ Ⓔ
29. Ⓐ Ⓑ Ⓒ Ⓓ Ⓔ
30. Ⓐ Ⓑ Ⓒ Ⓓ Ⓔ

31. Ⓐ Ⓑ Ⓒ Ⓓ Ⓔ
32. Ⓐ Ⓑ Ⓒ Ⓓ Ⓔ
33. Ⓐ Ⓑ Ⓒ Ⓓ Ⓔ
34. Ⓐ Ⓑ Ⓒ Ⓓ Ⓔ
35. Ⓐ Ⓑ Ⓒ Ⓓ Ⓔ
36. Ⓐ Ⓑ Ⓒ Ⓓ Ⓔ
37. Ⓐ Ⓑ Ⓒ Ⓓ Ⓔ
38. [＿＿＿＿＿＿]
39. Ⓐ Ⓑ Ⓒ Ⓓ Ⓔ
40. Ⓐ Ⓑ Ⓒ Ⓓ Ⓔ
41. Ⓐ Ⓑ Ⓒ Ⓓ Ⓔ
42. Ⓐ Ⓑ Ⓒ Ⓓ Ⓔ
43. Ⓐ Ⓑ Ⓒ Ⓓ Ⓔ
44. Ⓐ Ⓑ Ⓒ Ⓓ Ⓔ

PRACTICE TEST 2

Answer Sheet

1. Ⓐ Ⓑ Ⓒ Ⓓ Ⓔ	16. Ⓐ Ⓑ Ⓒ Ⓓ Ⓔ	31. Ⓐ Ⓑ Ⓒ Ⓓ Ⓔ
2. Ⓐ Ⓑ Ⓒ Ⓓ Ⓔ	17. Ⓐ Ⓑ Ⓒ Ⓓ Ⓔ	32. Ⓐ Ⓑ Ⓒ Ⓓ Ⓔ
3. Ⓐ Ⓑ Ⓒ Ⓓ Ⓔ	18. Ⓐ Ⓑ Ⓒ Ⓓ Ⓔ	33. Ⓐ Ⓑ Ⓒ Ⓓ Ⓔ
4. Ⓐ Ⓑ Ⓒ Ⓓ Ⓔ	19. Ⓐ Ⓑ Ⓒ Ⓓ Ⓔ	34. Ⓐ Ⓑ Ⓒ Ⓓ Ⓔ
5. Ⓐ Ⓑ Ⓒ Ⓓ Ⓔ	20. Ⓐ Ⓑ Ⓒ Ⓓ Ⓔ	35. Ⓐ Ⓑ Ⓒ Ⓓ Ⓔ
6. Ⓐ Ⓑ Ⓒ Ⓓ Ⓔ	21. Ⓐ Ⓑ Ⓒ Ⓓ Ⓔ	36. ▭
7. Ⓐ Ⓑ Ⓒ Ⓓ Ⓔ	22. Ⓐ Ⓑ Ⓒ Ⓓ Ⓔ	37. Ⓐ Ⓑ Ⓒ Ⓓ Ⓔ
8. Ⓐ Ⓑ Ⓒ Ⓓ Ⓔ	23. Ⓐ Ⓑ Ⓒ Ⓓ Ⓔ	38. Ⓐ Ⓑ Ⓒ Ⓓ Ⓔ
9. Ⓐ Ⓑ Ⓒ Ⓓ Ⓔ	24. Ⓐ Ⓑ Ⓒ Ⓓ Ⓔ	39. Ⓐ Ⓑ Ⓒ Ⓓ Ⓔ
10. Ⓐ Ⓑ Ⓒ Ⓓ Ⓔ	25. ▭	40. Ⓐ Ⓑ Ⓒ Ⓓ Ⓔ
11. Ⓐ Ⓑ Ⓒ Ⓓ Ⓔ	26. Ⓐ Ⓑ Ⓒ Ⓓ Ⓔ	41. Ⓐ Ⓑ Ⓒ Ⓓ Ⓔ
12. Ⓐ Ⓑ Ⓒ Ⓓ Ⓔ	27. Ⓐ Ⓑ Ⓒ Ⓓ Ⓔ	42. Ⓐ Ⓑ Ⓒ Ⓓ Ⓔ
13. Ⓐ Ⓑ Ⓒ Ⓓ Ⓔ	28. Ⓐ Ⓑ Ⓒ Ⓓ Ⓔ	43. Ⓐ Ⓑ Ⓒ Ⓓ Ⓔ
14. Ⓐ Ⓑ Ⓒ Ⓓ Ⓔ	29. Ⓐ Ⓑ Ⓒ Ⓓ Ⓔ	44. Ⓐ Ⓑ Ⓒ Ⓓ Ⓔ
15. ▭	30. Ⓐ Ⓑ Ⓒ Ⓓ Ⓔ	

Glossary

absolute maximum: the largest function value over its domain.

absolute minimum: the smallest function value over its domain.

acceleration: the change in velocity per unit time.

algebraic function: a function containing variables and constants whose exponents are any rational numbers; includes the categories of rational functions and polynomials.

antiderivative: the original function whose derivative is known.

area: the amount of square units enclosed by a given two-dimensional figure.

arithmetic sequence: a sequence for which there is a common difference between consecutive numbers.

average rate of change: for two specific points of a function, this is the quotient of the difference of the function values and the difference of the x values.

average value of a function: the numerical value of the integral of a function between an upper and lower boundary divided by the difference of the two boundaries.

average velocity: the ratio of the distance traveled to the elapsed time.

bounded region: a section of the xy-plane that is completely enclosed.

Cartesian plane: a surface that is divided into four equal sections, called quadrants, and for which each point is identified by its horizontal and vertical distance from the origin; also known as the xy-plane and the coordinate plane.

chain rule: a method for finding a derivative in which three variables are involved.

closed interval: an interval of numbers in which both the lowest and highest numbers listed are included.

common logarithm: an exponent for which the base is 10.

composite function: the result of a function rule applied to a second function.

compound interest: money earned on principal in which the given percent is applied to an increased principal for the next interest period.

concave down: a description of the portions of a curve in which the second derivative is decreasing.

concave up: a description of the portions of a curve in which the second derivative is increasing.

concavity: a measure that determines the change of the first derivative.

cone: a three-dimensional figure that has a circular base, a vertex directly above the center of the base, and a lateral surface that is curved.

constant: a quantity that has a specific numerical value.

continuity at a point: the property that the function can be drawn through this point with no breaks in the graph of the function.

cosecant of an angle: in a right triangle, the ratio of the hypotenuse to the opposite side.

cosine of an angle: in a right triangle, the ratio of the adjacent side to the hypotenuse.

cotangent of an angle: in a right triangle, the ratio of the adjacent side to the opposite side.

critical point: a point in the domain of a function for which the first derivative is either zero or undefined.

cusp: a point in the domain of a function for which the first derivatives on the left and right side are additive opposites.

cylinder: a three-dimensional figure that has two parallel and congruent circular bases and a lateral surface that is curved.

definite integral: an integral expression that contains lower and upper limits.

differential: a small incremental change in either x or y.

differential equation: an equation which contains the instantaneous change of one variable with respect to a second variable.

differentiation: the process of determining the derivative of a function.

discontinuity at a point: the property that the function cannot be drawn through this point without a break in the graph.

displacement: the difference between the final horizontal (vertical) position and the initial horizontal (vertical) position of a moving particle.

Extreme Value Theorem: a function that is continuous on a closed interval contains both a maximum and a minimum value at points of the interval.

first derivative: the instantaneous change of a function value at a specific point.

function: a relation in which for each value of the independent variable (usually x), there is only one value of the dependent variable (usually y).

geometric sequence: a sequence for which there is a common ratio between consecutive numbers.

half-open interval: an interval of numbers that includes only one of the lowest or highest number.

horizontal asymptote: a horizontal line that represents the limiting value of a function as x approaches infinity or negative infinity.

horizontal axis: the x-axis in the Cartesian plane.

implicit differentiation: the process of differentiation in which one variable is not expressed directly in terms of a second variable.

indefinite integral: an integral expression that contains neither lower nor upper limits.

infinity: denoted as ∞, it is an imaginary place on the number line that lies farther to the right from zero than any real number.

inflection point: a point at which the second derivative changes sign and for which a tangent line can be drawn.

instantaneous velocity: the limiting value of the average velocity in a small period of time.

integral: a symbol used to designate finding an antiderivative.

integrand: the function that appears next to the integral.

integration: the process of determining the antiderivative.

Intermediate Value Theorem: given a continuous function $f(x)$ and any closed interval $[a, b]$ there exists at least one x value between a and b that corresponds to a function value between $f(a)$ and $f(b)$.

inverse of a function $f(x)$: a second function, denoted as $f^{-1}(x)$, such that its graph is the reflection of $f(x)$ over the line $y = x$.

lateral area: the area of the non-base surface(s) of a three-dimensional figure.

limit: the value that a function $f(x)$ approaches as x approaches a specific number or infinity, or negative infinity.

linearization: the process by which the equation of a differentiable function at a given point is written in slope-intercept form.

local maximum: the largest value of a function over a specific interval.

local minimum: the smallest value of a function over a specific interval.

Mean Value Theorem for derivatives: Given a continuous function $f(x)$ on a closed interval, there exists at least one specific x value in this interval such that its derivative is equal to the slope of the line segment that connects the endpoints of the graph of $f(x)$.

Mean Value Theorem for definite integrals: given a continuous function $f(x)$ on a closed interval, there exists at least one specific x value in this interval such that its function value equals the value of the definite integral divided by the interval length.

natural logarithm: an exponent for which the base is e.

negative infinity: denoted as $-\infty$, it is an imaginary place on the number line that lies farther to the left from zero than any real number.

one-sided limit: the value that a function $f(x)$ approaches as x approaches a specific number either the left side or the right side.

open interval: an interval of numbers in which that highest and lowest numbers listed are not included.

optimization: the process by which derivatives are used to maximize or minimize a specific quantity such as area, cost, and volume.

origin: the point $(0,0)$ on the Cartesian plane.

oscillation: the behavior that describes a function that neither converges to a real number limit nor diverges to plus or minus infinity.

piecewise function: a function that is defined differently for at least two locations of the domain. A location may be a point or an interval.

point of tangency: the location at which a line crosses the graph of a given function at just one point.

polynomial function: a function containing variables and constants whose exponents are nonnegative integers.

pyramid: a three-dimensional figure whose base is a polygon and whose (lateral) faces are triangles that meet at a point (called the vertex).

rational function: a quotient of two polynomial functions.

Rectangular Approximation Method: a procedure that uses the areas of a series of connected rectangles to approximate the area between the graph of a function and the x-axis.

related rate: the change in a variable with respect to time.

Riemann Sum: the computed area between the graph of a function and the x-axis that uses either the Rectangular Approximation Method or the Trapezoidal Rule.

secant line: a line that intersects the graph of a function in at least two points.

secant of an angle: in a right triangle, the ratio of the hypotenuse to the adjacent side.

second derivative: the instantaneous change of the first derivative at a specific point.

sine of an angle: in a right triangle, the ratio of the opposite side to the hypotenuse.

slope: given two points on the graph of a linear function, it is the ratio of the difference of the y-values to the difference of the x-values.

speed: the ratio of distance per unit of time; it is also the absolute value of velocity.

sphere: a three-dimensional figure that represents the locus of all points at a given distance from a given point.

surface area: the combined area of the base(s) and the lateral surface(s) of a three-dimensional figure.

tangent line: the limit of a secant line between two points on the graph of a function, as the distance between the two points gets increasingly smaller.

tangent of an angle: in a right triangle, the ratio of the opposite side to the adjacent side.

transcendental function: a function that is not algebraic; examples include trigonometric and exponential functions.

Trapezoidal Rule: a procedure that uses the areas of a series of connected trapezoids to approximate the area between the graph of a function and the x-axis.

u-substitution: a process by which the integrand is made simpler to integrate; usually, the quantity $u\,du$ replaces the original quantity $x\,dx$.

two-sided limit: the single value that a function $f(x)$ approaches as x approaches a specific number from both the left side and the right side.

variable: a quantity that has not been assigned a specific numerical value.

velocity: speed in a specific direction.

vertical asymptote: given a function $f(x)$, a vertical line for which the function increases or decreases without bound as x approaches a specific real number.

vertical axis: the y-axis in the Cartesian plane.

volume: the amount of cubic units enclosed by a given three-dimensional figure.

x-intercept: the point of intersection of the graph of any curve and the x-axis.

y-intercept: the point of intersection of the graph of any curve and the y-axis.

zero of a function: the x value of the x-intercept, when the curve is a function.

Index

A

Absolute (global) extremes, 114
Acceleration, 136, 215–217
Antiderivative
  analytic rewriting of integrands, 174–175
  concept of, 167–169
  constant multiple property, 171
  definition of, 167
  formulas for, 169–170
  integral of sum property, 171
  by u-substitution, 171–173
Arccosecant, derivative of, 93
Arccosine, derivative of, 93
Arccotangent, derivative of, 93
Arcsecant, derivative of, 93
Arcsine, derivative of, 93
Arctangent, derivative of, 92–93
Area, 227–232
  bounded regions, 230–231
  under a curve
    numerical approximation, 175–176
    Rectangular Approximation Method, 177–181
    trapezoidal rule, 181–184
  summing two regions, 231–232
  between two curves, 227
  between two functions, 227–229
Average rate of change, 39–41
Average value, 222–226
Average velocity, 135

B

Bounded regions, 230–231

C

Chain rule, 81–88
  defined, 82
  higher-order derivatives with, 87–88
Leibniz notation form, 82
outside-inside principle, 82–85
Compounded interest, 219–220
Concavity, 124–126
  basic possible shapes of, 125
  defined, 124
  intervals of, 124–125
Constant
  constant times a function derivative, 69
  derivative of, 67
Constant multiple property, 171
Continuity
  of a function, 28–30
  Intermediate Value Theorem, 29–30
  at a point, 25–28
    implicit conditions of, 26
Corner
  differentiability and, 51
Cosecant function, derivative of, 76
Cosine function, derivative of, 76
Cotangent function
  derivative of, 76
Critical points, 112
Curve, area under
  numerical approximation, 175–176
  Rectangular Approximation Method, 177–181
Curve sketching, 131–134
Cusp
  differentiability and, 51

D

Decay, 217–219
Decreasing functions, 119–124
  definition of, 120
  increasing/decreasing derivative test, 120–121
Definite integrals
  as area, 185–186
  definition of, 185

Fundamental Theorem of Calculus, 190–195
Mean Value Theorem for, 224–226
properties of, 187–190
u-substitution, 195–197
Derivatives
applications of
first-derivative, 111–124
L'Hôpital's Rule, 149–152
linearization, 146–149
linear particle motion, 134–138
optimization, 138–143
related rates, 143–146
second-derivative, 124–134
differentiability of, 50–52
endpoints and, 52–53
as a function, 48–50
introduction to
of function at point $x = a$, 45–46
of function at point $x = a$ (x approaches a form), 46–48
linear approximations, 54–56
rates of change, 39–44
rules of
chain rule, 81–88
of constant, 67
of constant times a function, 69
exponential and logarithmic functions, 78–80
higher-order derivatives, 81, 87–88
implicit differentiation, 88–91
of inverse functions, 94–97
inverse trigonometric functions, 91–94
linear function, 68
outside-inside principle, 82–85
polynomials, 67–70
of a power, 68
products, 70–72
quotients, 72–74
of a sum, 69
trigonometric functions, 74–78
Differentiability of derivatives, 50–52
Differential equations, 211–222
compounded interest, 219–220
definition of, 211
distance and displacement, 220–222

exponential growth and decay, 217–219
with initial conditions, 213–215
position, velocity and acceleration, 215–217
separable, 211–213
Discontinuity
defined, 13
differentiability and, 51–52
infinite, 26, 28
jump, 26, 28
removable, 26–28
removing, 28
Displacement, 135, 220–222
Distance, 220–222

E

Endpoints
differentiability and, 52–53
Examining end behavior, 21
Exponential functions
derivatives of, 78–80
Exponential growth, 217–219
Extreme value theorem, 117

F

First-derivative applications
absolute extreme values, 114–117
first derivative test for local extremes, 123–124
intervals of increasing and decreasing functions, 119–124
local extreme values, 111–114
Mean Value Theorem, 117–119
Functions
area between two functions, 227–229
constant times a function derivative, 69
continuity of, 28–30
definition of increasing/decreasing, 120
derivative of, 48–50
inverse function derivative, 94–97
Fundamental Theorem of Calculus
Part 1, 190–193
Part 2, 194–195

H

Higher-order derivatives, 81
  with chain rule, 87–88
  notation for, 81

I

Implicit differentiation, 88–91
Increasing/decreasing derivative test,
    120–121
Increasing function, 119–124
  definition of, 120
  increasing/decreasing derivative test,
    120–121
Indefinite integrals
  constant multiple property, 171
  integral of sum property, 171
Infinite discontinuities, 26, 28
Infinity
  limits involving
    as x approaches a constant, 20–21
    as x approaches a positive or negative
      infinity, 20–21
Inflection points, 126–130
Initial conditions, 213–215
Instantaneous rate of change, 42–44
Instantaneous velocity, 135
Integral applications
  area, 227–232
  average value, 222–226
  differential equations, 211–222
Integral Evaluation, 191
Integral of sum property, 171
Integrands
  analytic rewriting of, 174–175
Interest, compounded, 219–220
Intermediate Value Theorem, 29–30
Inverse functions
  derivatives of, 94–97
Inverse trigonometric functions
  derivatives of, 91–94

J

Jump discontinuity, 26, 28

L

Leibniz notation form, 82
L'Hôpital's Rule, 149–152
Limits
  importance of, 13
  involving infinity
    as x approaches a constant, 20–21
    as x approaches a positive or negative
      infinity, 20–21
  properties of, 18
  as x approaches a constant
    algebraic approach, 18–20
    existence of limit, 15–17
    graphical approach, 14–17
    numerical approach, 14
    properties of limits, 18
    sandwich theorem, 17
Linear approximations of derivatives,
    54–56
Linear function
  derivative of, 68
Linearization, 146–149
Linear particle motion, 134–138
Local extreme values, 111–114
  critical point, 112
  first derivative test for, 123–124
  second derivative test for local extremes,
    131
Local linearity, 54
Logarithmic functions
  derivatives of, 78–80
LRAM, 177

M

Maximum values, 111–114
Mean Value Theorem, 117–119
  for definite integrals, 224–226
Minimum values, 111–114
MRAM, 177

N

Numerical approximation
  area under a curve, 175–176

Rectangular Approximation Method, 177–181
trapezoidal rule, 181–184

O

Optimization, 138–143
Outside-inside principle, 82–85

P

Point
    continuity at, 25–28
    slope at point, 44–48
Polynomials
    derivatives of, 67–70
Position, 215–217
Power
    derivative of, 68
Product Rule, 71
Products
    derivatives of, 70–72

Q

Quotient rule, 72
Quotients
    derivatives of, 72–74

R

Rates of change, 39–44
    average, 39–41
    instantaneous, 42–44
Rectangular Approximation Method, 177–181
Related rates, 143–146
Relative growth rate, 217
Removable discontinuities, 26–28
Riemann sum, 180–181, 184, 187, 191, 223
RRAM, 177

S

Sandwich theorem, 17

Secant function
    derivative of, 76
Second derivative
    applications
        concavity, 124–126
        curve sketching, 131–134
        inflection points, 126–130
        second-derivative test for local extremes, 131
    defined, 81
    notation for, 81
Sine function
    derivative of, 74–76
Slope
    at a point, 44–48
Speed, 136
Sum
    derivative of, 69

T

Tangent function
    derivative of, 76
Trapezoidal rule, 181–184
Trigonometric functions
    derivatives of, 74–78
    inverse
        derivatives of, 91–94

U

U-substitution
    antidifferentiation by, 171–173
    definitive integrals, 195–197

V

Velocity
    average, 135
    differential equations, 215–217
    instantaneous, 135
Vertical tangent
    differentiability and, 51

Notes

Notes

Notes

Notes

REA's Study Guides

Review Books, Refreshers, and Comprehensive References

Problem Solvers®

Presenting an answer to the pressing need for easy-to-understand and up-to-date study guides detailing the wide world of mathematics and science.

High School Tutors®

In-depth guides that cover the length and breadth of the science and math subjects taught in high schools nationwide.

Essentials®

An insightful series of more useful, more practical, and more informative references comprehensively covering more than 150 subjects.

Super Reviews®

Don't miss a thing! Review it all thoroughly with this series of complete subject references at an affordable price.

Interactive Flashcard Books®

Flip through these essential, interactive study aids that go far beyond ordinary flashcards.

Reference

Explore dozens of clearly written, practical guides covering a wide scope of subjects from business to engineering to languages and many more.

For information about any of REA's books, visit
www.rea.com

Research & Education Association
61 Ethel Road W., Piscataway, NJ 08854
Phone: (732) 819-8880

REA's Test Preps
The Best in Test Preparation

- REA "Test Preps" are *far more* comprehensive than any other test preparation series
- Each book contains full-length practice tests based on the most recent exams
- **Every** type of question likely to be given on the exams is included
- Answers are accompanied by **full** and **detailed** explanations

REA publishes hundreds of test prep books. Some of our titles include:

Advanced Placement Exams (APs)
Art History
Biology
Calculus AB & BC
Chemistry
Economics
English Language & Composition
English Literature & Composition
European History
French Language
Government & Politics
Latin Vergil
Physics B & C
Psychology
Spanish Language
Statistics
United States History
World History

College-Level Examination Program (CLEP)
American Government
College Algebra
General Examinations
History of the United States I
History of the United States II
Introduction to Educational Psychology
Human Growth and Development
Introductory Psychology
Introductory Sociology
Principles of Management
Principles of Marketing
Spanish
Western Civilization I
Western Civilization II

SAT Subject Tests
Biology E/M
Chemistry
French
German
Literature
Mathematics Level 1, 2
Physics
Spanish
United States History

Graduate Record Exams (GREs)
Biology
Chemistry
Computer Science
General
Literature in English
Mathematics
Physics
Psychology

ACT - ACT Assessment
ASVAB - Armed Services Vocational Aptitude Battery
CBEST - California Basic Educational Skills Test
CDL - Commercial Driver License Exam
COOP, HSPT & TACHS - Catholic High School Admission Tests
FE (EIT) - Fundamentals of Engineering Exams
FTCE - Florida Teacher Certification Examinations

GED
GMAT - Graduate Management Admission Test
LSAT - Law School Admission Test
MAT - Miller Analogies Test
MCAT - Medical College Admission Test
MTEL - Massachusetts Tests for Educator Licensure
NJ HSPA - New Jersey High School Proficiency Assessment
NYSTCE - New York State Teacher Certification Examinations
PRAXIS PLT - Principles of Learning & Teaching Tests
PRAXIS PPST - Pre-Professional Skills Tests
PSAT/NMSQT
SAT
TExES - Texas Examinations of Educator Standards
THEA - Texas Higher Education Assessment
TOEFL - Test of English as a Foreign Language
USMLE Steps 1,2,3 - U.S. Medical Licensing Exams

For information about any of REA's books, visit www.rea.com

Research & Education Association
61 Ethel Road W., Piscataway, NJ 08854
Phone: (732) 819-8880